4900 laCross Road
North Charleston SC, 29406, USA

Published by Createspace: 06/15/2015

ISBN-13: 978-1514329658
ISBN-10: 1514329654

BOOKS BY DR. STEVE WARNER

28 SAT Math Lessons to Improve Your Score in One Month
- Beginner Course
- Intermediate Course
- Advanced Course

New SAT Math Problems Arranged by Topic and Difficulty Level
320 SAT Math Problems Arranged by Topic and Difficulty Level
320 SAT Math Subject Test Problems Arranged by Topic and Difficulty Level
- Level 1 Test
- Level 2 Test

SAT Prep Book of Advanced Math Problems
The 32 Most Effective SAT Math Strategies
SAT Prep Official Study Guide Math Companion
320 ACT Math Problems Arranged by Topic and Difficulty Level
320 AP Calculus AB Problems Arranged by Topic and Difficulty Level
320 AP Calculus BC Problems Arranged by Topic and Difficulty Level
555 Math IQ Questions for Middle School Students
555 Geometry Problems for High School Students

BOOKS BY TAYYIP ORAL

555 Geometry Problems for High School Students
555 Math IQ Questions for Middle School Students
555 Math IQ Questions for Elementary School Students
IQ Intelligence Questions for Middle and High School Students
Master's Degree Program Preparation (IQ)
A Text Book for Job Placement Exam in Azerbaijan for Undergraduate and Post Undergraduate Students in Azerbaijan
Algebra (Text Book)
Geometry (Text book)
Geometry Formulas (Text Book)
Algebra Formulas (Text Book)

Algebra Handbook
for Gifted Middle School Students

Strategies, Concepts, and More Than 700 Algebra Problems with Solutions

---

Dr. Steve Warner, Tayyip Oral, Serife Turan

Table of Contents

Actions to Complete Before You Read This Book

1. Purchase a TI-84 or equivalent calculator

It is recommended that you use a TI-84 or comparable calculator for the Mathcounts exams. Answer explanations in this book will always assume you are using such a calculator.

2. Claim your FREE bonus

Visit the following webpage and enter your email address to receive additional problems with solutions and a free PDF version of *555 Geometry Problems for High School Students*.

www.thesatmathprep.com/McountX.html

1. INTEGERS

<u>Definition:</u> The set Z = {.........-5, -4, -3, -2, -1, 0, 1, 2, 3, 4, 5,........} is defined as the **set of integers**, and each element of this set is called an **integer**.

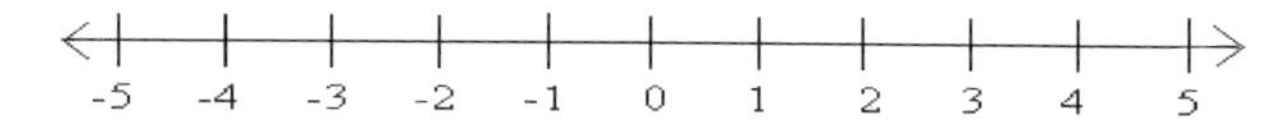

The set Z^{+} = {1, 2, 3, 4, 5, ...} is defined as the set of *positive* integers, or the set of **natural numbers**.

The set Z^{-} ={..........., -5, -4, -3, -2, -1} is defined as the set of *negative* integers.

If we include 0 with the set of positive integers, we get the **whole numbers**: W = {0, 1, 2, 3, 4, 5, ...}

OPERATIONS WITH INTEGERS

<u>Addition of Integers:</u>

In order to add two integers with opposite signs

- the integer with smaller absolute value is subtracted from the integer with larger absolute value, and
- the sign of the integer with larger absolute value is used for the result.

- To add -7 and +3, we first subtract 7 – 3 = 4, and then attach the "minus sign" because 7 is greater than 3. So the result is **-4**.
- Let us perform the operation (+3) + (+5), this time using the number line:

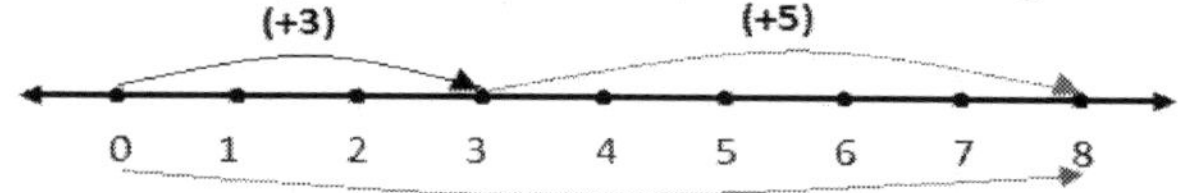

So (+3) + (+5) = +8

- Let us perform the operation (-4) + (+6), again using the number line:

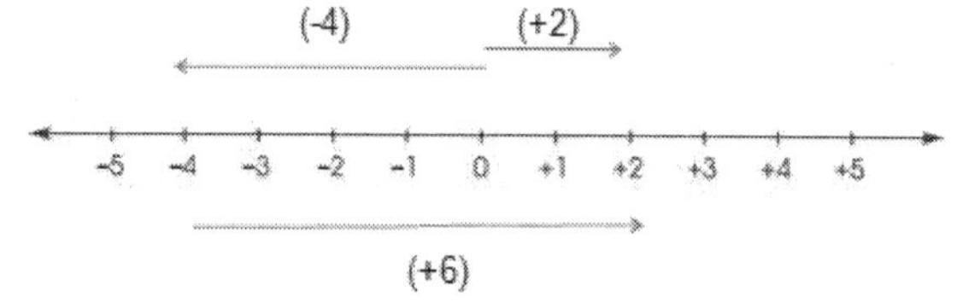

So (-4) + (+6) = +2

Notation: We use the symbol "$\in$" to indicate membership in a set. This symbol can be read as "is a member of" or "is in." For example, $2 \in Z$ can be read as "2 is in the set of integers," or more briefly as "2 is an integer."

Note that "$2 \in Z$" is a true statement, whereas "$0.2 \in Z$" is false. In the latter case we would write $0.2 \notin Z$.

Rule (commutative law): When $x, y \in Z$, $x + y = y + x$

Rule (associative law): When $x, y, z \in Z$, $x + (y + z) = (x + y) + z$

Rule (additive inverse): The *additive inverse* of the integer x is $-x$

- The additive inverse of the integer 6 is -6
- The additive inverse of -7 is 7 (note that $-(-7) = 7$)

Definition: $a - b = (a) + (-b)$

- $(+6) + (-3) = 6 - 3 = +3$
- $(+10) + (-4) = 10 - 4 = +6$
- $(+4) + (-9) = 4 - 9 = -5$
- $(-8) + (-8) = -8 - 8 = -16$
- $(-24) + (12) = -24 + 12 = -12$
- $(-36) + (+12) = -36 + 12 = -24$
- $(-8) + (+8) - (-4) = -8 + 8 + 4 = 4$
- $(0) - (-6) + (-6) + 20 = 0 + 6 - 6 + 20 = 20$

Multiplication of Integers:

When multiplying two integers with the same sign, the product will be a positive integer.

$(-) \cdot (-) = +$ $\qquad$ $(+) \cdot (+) = +$

- $(-3) \cdot (-4) = +12$
- $(-6) \cdot (-5) = +30$
- $(+5) \cdot (+2) = +10$
- $(+7) \cdot (+10) = +70$

When multiplying two integers with opposite signs, the product will be a negative integer.

$(-) \cdot (+) = -$ $\qquad$ $(+) \cdot (-) = -$

- $(-3) \cdot (+4) = -12$
- $(-5) \cdot (+3) = -15$
- $(+4) \cdot (-1) = -4$
- $(+2) \cdot (-7) = -14$

Example 1.1:

$(-3) \cdot (+3) \cdot (-2) = \mathbf{18}$ $\qquad$ $(-) \cdot (+) \cdot (-) = +$

$(-2) \cdot (+5) \cdot (-10) = \mathbf{100}$ $\qquad$ $(-) \cdot (+) \cdot (-) = +$

- Note that when a number is positive, we often leave off the + sign. For example 18 = +18.

Example 1.2:

Find the product of -6, -2, and -1.

Solution: $(-6) \cdot (-2) \cdot (-1) = \mathbf{-12}$.

- Note that the product of the three integers is negative because there are three negative *factors* (the factors are -6, -2, and -1).

Rule (commutative law): When $a, b \in Z$, $a \cdot b = b \cdot a$

- $3 \cdot 2 = 2 \cdot 3$ $\quad (+4) \cdot (-3) = (-3) \cdot (+4)$

Rule (associative law): When $a, b, c \in Z$, $(a \cdot b) \cdot c = a \cdot (b \cdot c)$

- $(3 \cdot 4) \cdot (-2) = 3 \cdot (4 \cdot (-2))$

Rule (identity element for multiplication is 1): When $a \in Z$, $a \cdot 1 = 1 \cdot a$

Rule (zero property): When $a \in Z$, $a \cdot 0 = 0$

Rule (multiplication is *distributive* over addition and subtraction):

When $a, b, c \in Z$, $a \cdot (b + c) = a \cdot b + a \cdot c$ and $a \cdot (b - c) = a \cdot b - a \cdot c$

- $(-4)\ [6 + (-3)] = (-4) \cdot (+6) + (-4) \cdot (-3) = -24 + 12 = -12$
- $(-3)\ [2 + (-5)] = (-3) \cdot (2) + (-3) \cdot (-5) = -6 + 15 = 9$

Example 1.3: $A = -6 + 6 \cdot 2 + 12 \div 3$, $\quad B = 7 + 12 \cdot 2 - 60 \div 5$, $\quad A + B = ?$

Solution:

$A = -6 + 12 + 4 = 6 + 4 = 10$, $\quad B = +7 + 24 - 12 = 31 - 12 = 19$

$A + B = 10 + 19 = \mathbf{29}$.

Order of Operations: A quick review

PEMDAS	
P	Parentheses
E	Exponentiation
M	Multiplication
D	Division
A	Addition
S	Subtraction

Note that multiplication and division have the same priority, and addition and subtraction have the same priority.

Example 1.4: $(-4) \cdot [(-3) + (12 \div 4)] = ?$

Solution:

$(-4)\ [-3 + 12 \div 4] = (-4) \cdot (-3) + (-4) \cdot (12 \div 4) = 12 + (-4) \cdot 3 = 12 - 12 = \mathbf{0}.$

Example 1.5: $(-3) \cdot (+3) \cdot (+4) \cdot (-4) = ?$

Solution:

$(-3) \cdot (+3) \cdot (+4) \cdot (-4) = (-9) \cdot (+4) \cdot (-4) = (-36) \cdot (-4) = +144 = \mathbf{144}.$

Example 1.6: $(-2^1 + 2^2) \cdot (-3^1 + 3^2) = ?$

Solution:

$(-2^1 + 2^2) \cdot (-3^1 + 3^2) = (-2 + 4) \cdot (-3 + 9) = (+2) \cdot (+6) = \mathbf{12}.$

Example 1.7: $(4^2 - 4^1) \cdot (5^2 - 5^1) \cdot (3^2 - 3^1) = ?$

Solution:

$(4^2 - 4^1) \cdot (5^2 - 5^1) \cdot (3^2 - 3^1) = (16 - 4) \cdot (25 - 5) \cdot (9 - 3) = 12 \cdot 20 \cdot 6 = \mathbf{1440}.$

Example 1.8: If $a = -2$, find $a^3 + a^2 + a + 2$.

Solution:

$a^3 + a^2 + a + 2 = (-2)^3 + (-2)^2 + (-2) + 2 = -8 + 4 - 2 + 2 = -4 - 2 + 2 = -6 + 2 = \mathbf{-4}.$

<u>Division of Integers:</u>

$C = A \div B$, A = Dividend, B = Divisor, C = Quotient

When dividing two integers with the same sign, the quotient will be positive.

$(-) \div (-) = +$ $\qquad$ $(+) \div (+) = +$

- $(-8) \div (-4) = 2$
- $24 \div 3 = 8$

When dividing two integers with opposite signs, the quotient will be negative.

$(+) \div (-) = -$ $\qquad$ $(-) \div (+) = -$

- $(10) \div (-2) = -5$
- $(-30) \div (+6) = -5$

Example 1.9: $(-30) \div (-5) + (-10) \div (+5) = ?$

Solution: $(-30) \div (-5) + (-10) \div (+5) = 6 - 2 = 4$

Example 1.10: $[(-10) \cdot (-12)] \div [(-48) \div (+16)] = ?$

Solution: $[(-10) \cdot (-12)] \div [(-48) \div (+16)] = 120 \div (-3) = -40$

Example 1.11: If $a = 3$ and $b = -2$, then $(a^3 - a^2) \div (b^2 + b) = ?$

Solution:

$$(a^3 - a^2) \div (b^2 + b) = (3^3 - 3^2) \div ((-2)^2 + (-2)) = (27 - 9) \div (4 - 2) = 18 \div 2 = 9$$

CONSECUTIVE INTEGERS

Consecutive integers are integers that follow each other in order. The difference between consecutive integers is 1. Here are two examples.

1, 2, 3	these are three consecutive integers
-3, -2, -1, 0, 1	these are five consecutive integers

In general, if x is an integer, then x, $x + 1$, $x + 2$, $x + 3$, … are consecutive integers.

Consecutive even integers are even integers that follow each other in order. The difference between consecutive even integers is 2. Here are two examples.

2, 4, 6	these are three consecutive even integers
-6, -4, -2, 0, 2	these are five consecutive even integers

In general, if x is an even integer, then x, $x + 2$, $x + 4$, $x + 6$,… are consecutive even integers.

Consecutive odd integers are odd integers that follow each other in order. The difference between consecutive odd integers is 2. Here are two examples.

1, 3, 5	these are three consecutive odd integers
-5, -3, -1, 1, 3	these are five consecutive odd integers

In general, if x is an odd integer, then x, $x + 2$, $x + 4$, $x + 6$,… are consecutive odd integers.

ODD AND EVEN INTEGERS

We will use the following abbreviations:

- O = Odd Integer
- E = Even Integer

Two integers are said to have the same **parity** if they are both even or both odd.

Addition and Subtraction Properties

- When we add or subtract two integers with the same parity the result is even.
- When we add or subtract two integers that do not have the same parity the result is odd.

These two properties are stated in more detail, with examples, below:

$1 \rightarrow O + O = E \Rightarrow 7 + 7 = 14$

$2 \rightarrow O - O = E \Rightarrow 17 - 3 = 14$

$3 \rightarrow O + E = O \Rightarrow 17 + 4 = 21$

$4 \rightarrow O - E = O \Rightarrow 19 - 4 = 15$

$5 \rightarrow E + O = O \Rightarrow 12 + 5 = 17$

$6 \rightarrow E - O = O \Rightarrow 8 - 1 = 7$

$7 \rightarrow E + E = E \Rightarrow 6 + 4 = 10$

$8 \rightarrow E - E = E \Rightarrow 20 - 6 = 14$

Multiplication Properties

- When we multiply an even integer by *any* integer the result is even.
- When we multiply two odd integers the result is odd.

These two properties are stated in more detail, with examples, below:

$1 \rightarrow E \times E = E \Rightarrow 6 \times 4 = 24$

$2 \rightarrow E \times O = E \Rightarrow 6 \times 5 = 30$

$2 \rightarrow O \times E = E \Rightarrow 3 \times 12 = 36$

$4 \rightarrow O \times O = O \Rightarrow 5 \times 7 = 35$

Examples 1.12: $2^5 + 3^9 \Rightarrow E + O = O$

$3^{16} + 5^{13} \Rightarrow O + O = E$

$5^{18} - 3^{17} + 2^{14} \Rightarrow O - O + E = E + E = E$

$5^{12} - 3^{12} \Rightarrow O - O = E$

Exponent Properties

* $O^n \rightarrow O \ (n \in N)$
* $E^n \rightarrow E \ (n \in N)$

Example 1.13: 3^6, $(-9)^{10}$, 121^{31}, and 30^0 are all odd integers

➢ Note that the last expression in example 1.13 is quite different from the others. When we raise *any* nonzero number to the 0 power, the result is always 1 which is an odd integer.

Example 1.14: 8^4, -16^8, 204^{15}, and 24^9 are all even integers.

EVEN INTEGERS

An integer is **even** if it can be written in the form $2n$ for some integer n.

Examples 1.15: 2 is even because it can be written as $2 \cdot 1$.

8 is even because it can be written as $2 \cdot 4$.

0 is even because it can be written as $2 \cdot 0$.

-6 is even because it can be written as $2 \cdot (-3)$.

- An even integer is also said to be **divisible by 2.**
- The units digit of an even integer is 0, 2, 4, 6, or 8.
- When expressing an even integer as $2n$, it is important that n be an integer.

Example 1.16: 1 can be written as $2 \cdot (\frac{1}{2})$. This does *not* show that 1 is even because $\frac{1}{2}$ is *not* an integer.

Since $2 \cdot 0 = 0$ and $2 \cdot 1 = 2$, we see that it is impossible to write 1 in the form $2n$ where n is an integer. So 1 is *not* even.

Example 1.17: The sum of two consecutive even integers is 22. Find these integers.

Solution 1: Let us call the first even integer $2n$. It follows that the second even integer is $2n + 2$. We are given that $2n + (2n + 2) = 22$. So we have $4n + 2 = 22$. Therefore $4n = 22 - 2 = 20$. So $n = \frac{20}{4} = 5$.

First integer = $2n = 2 \cdot 5 = \mathbf{10}$. Second integer = $2n + 2 = 2 \cdot 5 + 2 = \mathbf{12}$.

Solution 2: Let us call the first even integer k. It follows that the second even integer is $k + 2$. We are given that $k + (k + 2) = 22$. So we have $2k + 2 = 22$. Therefore $2k = 22 - 2 = 20$. So $k = \frac{20}{2} = 10$. So the first integer is $k = \mathbf{10}$ and the second is $k + 2 = \mathbf{12}$.

Solution 3: If we take a guess that the first even integer is 8, it follows that the second even integer is 10. Since 8 + 10 = 18 < 22, our original guess was a bit too small.

So let's try 10 for the first integer. It follows that the second integer is 12. Since 10 + 12 = 22, the two integers are **10** and **12**.

Example 1.18: The sum of three consecutive even integers is 42. Find the result when we add all of the digits of these three integers.

Solution: First we will find the three integers. Let us call the first even integer $2n$. It follows that the second even integer is $2n + 2$ and the third even integer is $2n + 4$. We are given that $2n + (2n + 2) + (2n + 4) = 42$. So we have $6n + 6 = 42$. Therefore $6n = 42 - 6 = 36$. So $n = \frac{36}{6} = 6$.

First integer = $2n = 2 \cdot 6 = 12$.

Second integer = $2n + 2 = 2 \cdot 6 + 2 = 14$.

Third integer = $2n + 4 = 2 \cdot 6 + 4 = 16$.

The digits add up to 1 + 2 + 1 + 4 + 1 + 6 = **15**.

Remarks: (1) Instead of finding the three integers algebraically we can also find them by guessing as we did in solution 3 from example 1.17.

(2) A quick way to find the middle integer is to simply divide 42 by 3 to get $\frac{42}{3} = 14$. It then immediately follows that the three integers are 12, 14, and 16.

The reason that this works is because the **median** of a sequence of consecutive even integers is equal to the **arithmetic mean** of the sequence.

The **median** of a list of numbers is the middle number when the numbers are written in increasing (or decreasing) order. In this example, the median of 12, 14, and 16 is 14.

The **arithmetic mean** of a list of numbers is the sum of the numbers divided by the quantity of numbers in the list. In this example, the arithmetic mean of 12, 14, and 16 is $\frac{12+14+16}{3} = \frac{42}{3} = 14$.

Example 1.19: The product of two positive consecutive even integers is 48. Find the arithmetic and geometric means of these integers.

Solution: Let us call the first even integer $2n$. It follows that the second even integer is $2n + 2$. We are given that $2n(2n + 2) = 48$. So we have $4n^2 + 4n = 48$. Dividing each term by 4 gives us $n^2 + n = 12$. Subtracting 12 gives us $n^2 + n - 12 = 0$. So we have $(n + 4)(n - 3) = 0$. Therefore $n + 4 = 0$ or $n - 3 = 0$. So $n = -4$ or $n = 3$. Since we are given that the two integers are positive, we can reject -4, and we have that $n = 3$.

First integer = $2n = 2 \cdot 3 = 6$. Second integer = $2n + 2 = 2 \cdot 3 + 2 = 8$.

Arithmetic mean = $\frac{6+8}{2} = \frac{14}{2} = \mathbf{7}$.

Geometric mean = $\sqrt{6 \cdot 8} = \sqrt{48} = \sqrt{16 \cdot 3} = \sqrt{16}\sqrt{3} = \mathbf{4\sqrt{3}}$.

Remarks: (1) Instead of finding the two integers algebraically we can also find them by guessing as we did in solution 3 from example 1.17.

(2) The **geometric mean** of two numbers is the square root of the product of the two numbers. In this example, the geometric mean of 6 and 8 is $\sqrt{6 \cdot 8} = 4\sqrt{3}$.

Example 1.20: The sum of two consecutive even integers is 26. Find the ratio of the larger integer to the smaller integer.

Solution: Using one of the methods of solution from example 1.17 we see that the two integers are 12 and 14. The ratio of the larger integer to the smaller integer is then $\frac{14}{12} = \frac{7}{6}$.

Example 1.21: The positive difference between the squares of two consecutive even integers is 28. Find the product of these integers.

Solution: Let us call the first even integer $2n$. It follows that the second even integer is $2n + 2$. We are given that $(2n + 2)^2 - (2n)^2 = 28$. So we have $(2n + 2 - 2n)(2n + 2 + 2n) = 28$, or equivalently $2(4n + 2) = 28$. Dividing each side of this equation by 2 gives us $4n + 2 = 14$. Subtracting 2 gives us $4n = 14 - 2 = 12$. So we have $n = \frac{12}{4} = 3$.

First integer = $2n = 2 \cdot 3 = 6$. Second integer = $2n + 2 = 2 \cdot 3 + 2 = 8$.

The product of these integers is $6 \cdot 8 = \mathbf{48}$.

Remark: Instead of finding the two integers algebraically we can also find them by guessing as we did in solution 3 from example 1.17.

ODD INTEGERS

An integer is **odd** if it can be written in the form $2n + 1$ for some integer n.

Examples 1.22: 3 is odd because it can be written as $2 \cdot 1 + 1$.

7 is odd because it can be written as $2 \cdot 3 + 1$.

-5 is odd because it can be written as $2 \cdot (-3) + 1$.

- An odd integer is *not* divisible by 2.
- The units digit of an odd integer is 1, 3, 5, 7, or 9.
- When expressing an odd integer as $2n + 1$, it is important that n is an integer.

Example 1.23: 2 can be written as $2 \cdot (\frac{1}{2}) + 1$. This does *not* show that 2 is odd because $\frac{1}{2}$ is *not* an integer.

In fact, we have already shown that 2 is even ($2 = 2 \cdot 1$).

Example 1.24: The sum of two consecutive odd integers is 12. Find these integers.

Solution: Let us call the first odd integer k. It follows that the second odd integer is $k+2$. We are given that $k+(k+2)=12$. So $2k+2=12$. Therefore $2k=12-2=10$. So $k=\frac{10}{2}=5$. So the first integer is $k=\mathbf{5}$ and the second is $k+2=\mathbf{7}$.

Remark: Instead of finding the two integers algebraically we can also find them by guessing as we did in solution 3 from example 1.17.

Example 1.25: The sum of three consecutive odd integers is 15. Evaluate the product of these integers.

Solution: Let us call the first odd integer k. It follows that the second odd integer is $k+2$ and the third odd integer is $k+4$. We are given that $k+(k+2)+(k+4)=15$. So $3k+6=15$. Therefore $3k=15-6=9$. So $k=\frac{9}{3}=3$. So the first integer is $k=3$, the second is $k+2=5$, and the third integer is $k+4=7$. The product of these integers is $3 \cdot 5 \cdot 7 = \mathbf{105}$.

Remark: Instead of finding the three integers algebraically we can also find them by guessing as we did in solution 3 from example 1.17.

Example 1.26: The positive difference between the squares of two consecutive odd integers is 24. Find the values of these integers.

Solution: Let us call the first odd integer k. It follows that the second odd integer is $k+2$. We are given that $(k+2)^2-k^2=24$. So we have $(k+2-k)(k+2+k)=24$, or equivalently $2(2k+2)=24$. Dividing each side of this equation by 2 gives us $2k+2=12$. Subtracting 2 gives us $2k=12-2=10$. So we have $k=\frac{10}{2}=5$.

First integer = $k=\mathbf{5}$. Second integer = $k+2=5+2=\mathbf{7}$.

Example 1.27: The product of two positive consecutive odd integers is 35. Find the arithmetic and geometric means of these integers.

Solution: Let us call the first odd integer k. It follows that the second odd integer is $k+2$. We are given that $k(k+2)=35$. So $k^2+2k=35$. Subtracting 35 gives us $k^2+2k-35=0$. So we have $(k+7)(k-5)=0$. Therefore $k+7=0$ or $k-5=0$. So $k=-7$ or $k=5$. Since we are given that the two integers are positive, we can reject -7, and we have that $k=5$.

First integer = $k = 5$. Second integer = $k + 2 = 5 + 2 = 7$.

Arithmetic mean = $\frac{5+7}{2} = \frac{12}{2} = \mathbf{6}$.

Geometric mean = $\sqrt{5 \cdot 7} = \sqrt{\mathbf{35}}$.

Remarks: (1) Instead of finding the two integers algebraically we can also find them by guessing as we did in solution 3 from example 1.17.

(2) We will see the arithmetic and geometric means in more detail later.

Example 1.28: If the sum of three consecutive odd integers is b, describe the median of the three integers in terms of b.

Solution: Let us call the first odd integer k. It follows that the second odd integer is $k + 2$ and the third odd integer is $k + 4$. We are given that $k + (k + 2) + (k + 4) = b$. So $3k + 6 = b$. Therefore $3k = b - 6$. So we have $k = \frac{b-6}{3}$.

The median of the three integers is $k + 2 = \frac{b-6}{3} + 2 = \frac{b-6}{3} + \frac{6}{3} = \frac{\mathbf{b}}{\mathbf{3}}$.

Example 1.29: The sum of two consecutive odd integers is 28. Find the product of these integers.

Solution: Let us call the first odd integer k. It follows that the second odd integer is $k + 2$. We are given that $k + (k + 2) = 28$. So $2k + 2 = 28$. Therefore $2k = 28 - 2 = 26$. So $k = \frac{26}{2} = 13$. So the first integer is $k = 13$ and the second is $k + 2 = 15$.

It follows that the product is $13 \cdot 15 = \mathbf{195}$.

Example 1.30: If a is the sum of the first five positive consecutive odd integers, and b is the sum of the positive consecutive odd integers between 10 and 20, then evaluate $\frac{a}{b}$.

Solution: $a = 1 + 3 + 5 + 7 + 9 = 25$, $b = 11 + 13 + 15 + 17 + 19 = 75$.

So $\frac{a}{b} = \frac{25}{75} = \frac{\mathbf{1}}{\mathbf{3}}$.

FACTORIALS

The **factorial** of a positive integer n, written $n!$, is the product of all positive integers less than or equal to n.

$$n! = 1 \cdot 2 \cdot 3 \cdots n$$

$0!$ is defined to be 1, so that $n!$ is defined for all nonnegative integers n.

$0! = 1$

$1! = 1$

$2! = 1 \cdot 2 = 2$

$3! = 1 \cdot 2 \cdot 3 = 6$

$4! = 1 \cdot 2 \cdot 3 \cdot 4 = 24$

Example 1.31: Evaluate $2! + 3!$

Solution 1: $2! = 1 \cdot 2 = 2$ and $3! = 1 \cdot 2 \cdot 3 = 6$. So $2! + 3! = 2 + 6 = \mathbf{8}$.

Solution 2: $2! + 3! = 2! + 3 \cdot 2! = 2!\,(1 + 3) = 2!\,(4) = 2 \cdot 4 = \mathbf{8}$.

Example 1.32: $3! + 4! =$

Solution: $3! + 4! = 6 + 24 = \mathbf{30}$.

Example 1.33: Evaluate $\frac{10!+9!}{8!}$.

Solution: $\frac{10!+9!}{8!} = \frac{10\cdot9\cdot8!+9\cdot8!}{8!} = \frac{8!(90+9)}{8!} = \mathbf{99}$.

Example 1.34: Evaluate $\frac{7!+6!}{5!+4!}$

Solution: $\frac{7!+6!}{5!+4!} = \frac{7\cdot6\cdot5\cdot4!+6\cdot5\cdot4!}{5\cdot4!+4!} = \frac{4!(7\cdot6\cdot5+6\cdot5)}{4!(5+1)} = \frac{210+30}{6} = \frac{240}{6} = \mathbf{40}$.

Example 1.35: Evaluate $\frac{3\cdot6!-3\cdot5!}{4!-3!}$

Solution: $\frac{3\cdot6!-3\cdot5!}{4!-3!} = \frac{3\cdot6\cdot5!-3\cdot5!}{4\cdot3!-3!} = \frac{5!(3\cdot6-3)}{3!(4-1)} = \frac{5\cdot4(18-3)}{3} = \frac{20\cdot15}{3} = \mathbf{100}$.

Example 1.36: $0! + 3! + 4! =$

Solution: $0! + 3! + 4! = 1 + 3! + 4 \cdot 3! = 1 + 3!\,(1 + 4) = 1 + 6(5) = \mathbf{31}$.

Example 1.37: $5! - 4! + 2! - 1! =$

Solution: $5! - 4! + 2! - 1! = 5 \cdot 4! - 4! + 2 - 1 = 4!\,(5 - 1) + 1$

$= 24(4) + 1 = 96 + 1 = \mathbf{97}$.

Example 1.38: $\frac{4!}{0!} + \frac{3!}{1!} + \frac{2!}{2} =$

Solution: $\frac{4!}{0!} + \frac{3!}{1!} + \frac{2!}{2} = \frac{4\cdot3\cdot2\cdot1}{1} + \frac{3\cdot2\cdot1}{1} + \frac{2\cdot1}{2} = 24 + 6 + 1 = \mathbf{31}$.

Example 1.39: $\frac{10!}{8!} + \frac{9!}{8!} + \frac{8!}{7!} =$

Solution: $\frac{10!}{8!} + \frac{9!}{8!} + \frac{8!}{7!} = \frac{10\cdot9\cdot8!}{8!} + \frac{9\cdot8!}{8!} + \frac{8\cdot7!}{7!} = 90 + 9 + 8 = \mathbf{107}$.

Example 1.40: Evaluate $\frac{6!}{4!-3!}$.

Solution: $\frac{6!}{4!-3!} = \frac{6\cdot5\cdot4\cdot3!}{4\cdot3!-3!} = \frac{3!(6\cdot5\cdot4)}{3!(4-1)} = \frac{120}{3} = \mathbf{40}$.

Example 1.41: $\frac{9!+10!}{9!} =$

Solution: $\frac{9!+10!}{9!} = \frac{9!+10\cdot 9!}{9!} = \frac{9!(1+10)}{9!} = \mathbf{11}$.

Example 1.42: Evaluate $\frac{8!}{5!} - \frac{6!}{3!}$.

Solution: $\frac{8!}{5!} - \frac{6!}{3!} = \frac{8\cdot 7\cdot 6\cdot 5!}{5!} - \frac{6\cdot 5\cdot 4\cdot 3!}{3!} = 8\cdot 7\cdot 6 - 6\cdot 5\cdot 4 = 336 - 120 = \mathbf{216}$.

Example 1.43: Compute $13! - 12!$.

Solution: $13! - 12! = 13\cdot 12! - 12! = 12!\,(13-1) = \mathbf{12!\cdot 12}$.

Example 1.44: $\frac{10!+9!}{8!+7!} =$

Solution: $\frac{10!+9!}{8!+7!} = \frac{10\cdot 9!+9!}{8\cdot 7!+7!} = \frac{9!(10+1)}{7!(8+1)} = \frac{9\cdot 8(11)}{9} = 8\cdot 11 = \mathbf{88}$.

2. NUMBER THEORY

PRIME NUMBERS

Definition: An integer n is **divisible** by an integer d if there is another integer k such that $n = dk$.

For example, 42 is divisible by 7 because $42 = 7\cdot 6$.

In practice we can check if n is divisible by d simply by dividing n by d in our calculator. If the answer is an integer, then n is divisible by d. If the answer is not an integer (it contains digits after the decimal point), then n is not divisible by d.

Definition: If n is divisible by d, we say that d is a **divisor** of n (or d is a **factor** of n).

Definition: A **prime number** is a positive integer with *exactly* two divisors (1 and itself).

Here is a list of the first few primes:

2, 3, 5, 7, 11, 13, 17, 19, 23, 29, 31,...

Note that 1 is **not** prime. It only has one factor!

* The smallest prime number is 2.
* Negative integers are not included in the prime numbers.
* 2 is the only even prime number.

Example 2.1: Find the sum of the first five prime numbers.

Solution: The first five prime numbers are: 2, 3, 5, 7 and 11. The sum of these numbers is $2 + 3 + 5 + 7 + 11 = \mathbf{28}$.

Example 2.2: Find the result when we add the digits of the greatest two digit prime number.

Solution: The greatest two digit prime number is 97, and $9 + 7 = \mathbf{16}$.

Example 2.3: The product of two prime numbers is 22. Find the positive difference between these two numbers.

Solution: $22 = 2 \cdot 11$, and 2 and 11 are prime. So we have $11 - 2 = \mathbf{9}$.

Example 2.4: Find the product of all single digit prime numbers.

Solution: The single digit prime numbers are: 2, 3, 5 and 7. The product of these numbers is $2 \times 3 \times 5 \times 7 = \mathbf{210}$.

Example 2.5: Evaluate the sum of the prime numbers between 10 and 30.

Solution: The prime numbers between 10 and 30 are: 11, 13, 17, 19, 23, and 29. The sum of these numbers is $11 + 13 + 17 + 19 + 23 + 29 = \mathbf{112}$.

PRIME FACTORIZATION

The Fundamental Theorem of Arithmetic: Every integer greater than 1 can be written "uniquely" as a product of primes.

The word "uniquely" is written in quotes because prime factorizations are only unique if we agree to write the primes in increasing order.

For example, 6 can be written as $2 \cdot 3$ or as $3 \cdot 2$. But these two factorizations are the same except that we changed the order of the factors. To make things as simple as possible we always agree to use the **canonical representation**. The word "canonical" is just a fancy name for "natural," and the most natural way to write a prime factorization is in increasing order of primes. So the canonical representation of 6 is $2 \cdot 3$.

As another example, the canonical representation of 18 is $2 \cdot 3 \cdot 3$. We can tidy this up a bit by rewriting $3 \cdot 3$ as 3^2. So the canonical representation of 18 is $2 \cdot 3^2$.

If you are new to factoring, you may find it helpful to draw a factor tree. For example here is a factor tree for 18:

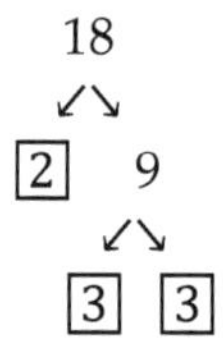

To draw this tree we started by writing 18 as the product $2 \cdot 9$. We put a box around 2 because 2 is prime, and does not need to be factored anymore. We then proceeded to factor 9 as $3 \cdot 3$. We put a box around each 3 because 3 is prime. We now see that we are done, and the prime factorization can be found by multiplying all of the boxed numbers together. Remember that we will usually want the canonical representation, so write the final product in increasing order of primes.

By the Fundamental Theorem of Arithmetic above it does not matter how we factor the number – we will always get the same canonical form. For example, here is a different factor tree for 18:

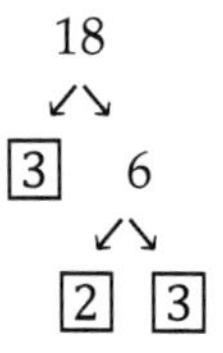

Here is an alternate way to find the prime factorization of 18:

18	2
9	3
3	3
1	

In the above table we started by placing the positive integer 18 in the upper left corner of the table. We then started with the smallest prime 2. Since 18 is divisible by 2, we write 2 the right of 18, and $\frac{18}{2} = 9$ below 18. Since 9 is not divisible by 2, we move to the next prime 3.

We repeat this process until we get a 1 on the left hand side. At this point we stop. The prime factorization of the number is then the product of the primes in the right hand column. In this case $18 = 2 \cdot 3 \cdot 3 = 2 \cdot 3^2$.

Example 2.6: Find the prime factors of 48.

Solution:

48	2
24	2
12	2
6	2
3	3
1	

$48 = 2^4 \cdot 3^1$. So the prime factors of 48 are **2** and **3**.

Example 2.7: How many prime factors does 280 have?

Solution:

280	2
140	2
70	2
35	5
7	7
1	

$280 = 2^3 \cdot 5^1 \cdot 7^1$. So the prime factors of 280 are 2, 5 and 7. Therefore 280 has **3** prime factors.

Example 2.8: Find the sum of the prime factors of 900.

Solution:

900	2
450	2
225	3
75	3
25	5
5	5
1	

$900 = 2^2 \cdot 3^2 \cdot 5^2$. So the prime factors of 900 are 2, 3, and 5. The sum of these prime factors is 2 + 3 + 5 = **10**.

Example 2.9: Find the prime factors of 17.

Solution: Since 17 is prime, the only prime factor of 17 is **17**.

Example 2.10: Find the prime factorization of 60 (in canonical form).

Solution:

60	2
30	2
15	3
5	5
1	

$60 = \mathbf{2^2 \cdot 3^1 \cdot 5^1}$.

FINDING ALL POSITIVE DIVISORS OF AN INTEGER

Let $A = a^m \cdot b^n \cdot c^k$ be the prime factorization of A. So a, b, and c are distinct prime numbers, and m, n, and k are positive integers. The number of positive divisors (or factors) of A is

$$T = (m+1)(n+1)(k+1).$$

Example 2.11: How many positive divisors does 24 have? Find them.

Solution: $24 = 3 \cdot 8 = 3 \cdot 2^3 = 2^3 \cdot 3^1$. So $m = 3$, $n = 1$ (and $k = 0$). So $T = (3+1)(1+1) = 4 \cdot 2 = \mathbf{8}$.

So 24 has 8 divisors. They are **1, 2, 3, 4, 6, 8, 12, and 24.**

Example 2.12: Find all positive divisors of 90.

Solution:

90	2
45	3
15	3
5	5
1	

So $90 = 2^1 \cdot 3^2 \cdot 5^1$, and 90 has $(1+1)(2+1)(1+1) = 2 \cdot 3 \cdot 2 = 12$ positive divisors. They are **1, 2, 3, 5, 6, 9, 10, 15, 18, 30, 45, and 90**.

Example 2.13: How many divisors (positive or negative) does 80 have?

Solution:

80	2
40	2
20	2
10	2
5	5
1	

So $80 = 2^4 \cdot 5^1$, and 80 has $(4+1)(1+1) = 5 \cdot 2 = 10$ positive divisors. Therefore the total number of divisors of 80 is $2 \cdot 10 =$ **20**.

Notes: (1) The number of negative divisors of an integer is equal to the number of positive divisors of an integer.

(2) The divisors of 80 are $\pm 1, \pm 2, \pm 4, \pm 5, \pm 8, \pm 10, \pm 16, \pm 20, \pm 40, \pm 80$.

Example 2.14: Find the number of positive divisors of 40.

Solution:

40	2
20	2
10	2
5	5
1	

So $40 = 2^3 \cdot 5^1$, and 40 has $(3+1)(1+1) = 4 \cdot 2 = \mathbf{8}$ positive divisors.

- The sum of the positive divisors and negative divisors of an integer is 0.

For example, the positive divisors of 40 are 1, 2, 4, 5, 8, 10, 20, and 40, and the negative divisors of 40 are -1, -2, -4, -5, -8, -10, -20, and -40.

Clearly the sum of all these numbers is 0.

Example 2.15: Find the number of divisors of 70.

Solution:

70	2
35	5
7	7
1	

$70 = 2^1 \cdot 5^1 \cdot 7^1$. So the number of positive divisors of 70 is

$(1+1)(1+1)(1+1) = 2 \cdot 2 \cdot 2 = 8$.

The number of negative divisors is also 8. So the total number of divisors of 70 is **16**.

Example 2.16: Find the number of divisors of 1600.

Solution:

1600	2
800	2
400	2
200	2
100	2
50	2
25	5
5	5
1	

$1600 = 2^6 \cdot 5^2$. So the number of positive divisors of 1600 is

$(6+1)(2+1) = 7 \cdot 3 = 21$.

The number of negative divisors is also 21. So the total number of divisors of 1600 is **42**.

DIVISIBILITY TRICKS

- **Divisibility by 2:** An integer is divisible by 2 precisely when the last digit is 0, 2, 4, 6 or 8.

An integer that is divisible by 2 is usually referred to as an even integer.

Example 2.17: 6, 38, 120, 414, and 3412 are all divisible by 2 (or even).

7, 13, and 415 are not divisible by 2 (or odd).

- **Divisibility by 3:** An integer is divisible by 3 precisely when the sum of its digits is divisible by 3.

Example 2.18: Is 6348 divisible by 3?

Solution: $6 + 3 + 4 + 8 = 21$. Since 21 is divisible by 3, so is 6348.

Example 2.19: Is 7,864,392 divisible by 3?

Solution: 7 + 8 + 6 + 4 + 3 + 9 + 2 = 39. Since 39 is divisible by 3, so is 7,864,392.

Example 2.20: Is 5215 divisible by 3?

Solution: 5 + 2 + 1 + 5 = 13. Since 13 is not divisible by 3, neither is 5215.

- **Divisibility by 4:** An integer is divisible by 4 precisely when the number formed by taking just the last two digits of the integer is divisible by 4.

Example 2.21: 4816, 120, 208, 1956, 9988, 9644 are all divisible by 4 because the number formed by taking the last two digits of each number (16, 20, 08, 56, 88, 44) is divisible by 4.

Example 2.22: 4815 is not divisible by 4 because 15 is not divisible by 4.

∗ The remainder when dividing a positive integer by 4 is equal to the remainder when dividing the number formed by the last 2 digits by 4.

Example 2.23: What is the remainder when 648,387 is divided by 4?

Solution: The remainder when dividing 648,387 by 4 is equal to the remainder when dividing 87 by 4. Now, 84 is divisible by 4. So when 87 is divided by 4, the remainder is **3**.

- **Divisibility by 5:** An integer is divisible by 5 precisely when the last digit of the integer is 0 or 5.

Example 2.24: 40, 120, 665, 8875, and 9900 are all divisible by 5.

41, 123, 8877, and 9909 are not divisible by 5.

∗ The remainder when dividing a positive integer by 5 is equal to the remainder when dividing the last digit of the integer by 5.

Example 2.25: What is the remainder when 9,986,978 is divided by 5?

Solution: The remainder when dividing 9,986,978 by 5 is equal to the remainder when dividing 8 by 5. Now, 5 is divisible by 5. So when 8 is divided by 5, the remainder is 8 – 5 = **3**.

Example 2.26: Find the remainder when 64,389 is divided by 5.

Solution: The remainder when dividing 64,389 by 5 is equal to the remainder when dividing 9 by 5. Now, 5 is divisible by 5. So when 9 is divided by 5, the remainder is 9 – 5 = **4**.

- **Divisibility by 6:** An integer is divisible by 6 if it is divisible by both 2 and 3.

Example 2.27: Is 98,642,820 divisible by 6?

Solution: Since 98,642,820 ends in 0, it is divisible by 2.

$9 + 8 + 6 + 4 + 2 + 8 + 2 + 0 = 39$. Since 39 is divisible by 3, so is 98,642,820.

Since 98,642,820 is divisible by both 2 and 3, it is divisible by 6.

- **Divisibility by 9:** An integer is divisible by 9 precisely when the sum of its digits is divisible by 9.

Example 2.28: Is 4986 divisible by 9?

Solution: $4 + 9 + 8 + 6 = 27$. Since 27 is divisible by 9, so is 4986.

* The remainder when dividing a positive integer by 9 is equal to the remainder when the sum of the digits is divided by 9.

Example 2.29: What is the remainder when the 12 digit integer 777,777,777,777 is divided by 9?

Solution: $12 \cdot 7 = 84$ (adding 7 to itself 12 times is the same as multiplying 7 by 12). Since the remainder when 84 is divided by 9 is 3, the remainder when 777,777,777,777 is divided by 9 is also 3.

- **Divisibility by 11:** Add up every other digit beginning with the second digit from the left, and then subtract the remaining digits. If this new number is divisible by 11, so is the original number. Otherwise it is not.

Example 2.30: Is 97,642,160 divisible by 11?

Solution: $(7 + 4 + 1 + 0) - (9 + 6 + 2 + 6) = 12 - 23 = -11$. Since -11 is divisible by 11, so is 97,642,160.

GCD and LCM

The **greatest common divisor (GCD)** (also known as the **greatest common factor (GCF)** or **highest common factor (HCF)**) of two or more positive integers is the largest positive integer that each integer in the set is divisible by.

The **least common multiple (LCM)** of two or more positive integers is the smallest positive integer that is divisible by each integer in the set.

Example 2.31: Find the GCD and LCM of 9 and 15.

Solution 1: The factors of 9 are 1, 3 and 9. The factors of 15 are 1, 3, 5 and 15. So the common factors of 9 and 15 are 1 and 3. So GCD(9,15) = **3**.

The multiples of 9 are 9, 18, 27, 36, 45, 54, 63,... and the multiples of 15 are 15, 30, 45,.. We can stop at 45 because 45 is also a multiple of 9. So LCM(9,15) = **45**.

Solution 2: The prime factorizations of 9 and 15 are $9 = 3^2$ and $15 = 3 \cdot 5$. To find the GCD we multiply together the smallest powers of each prime from both factorizations, and for the LCM we multiply the highest powers of each prime. So GCD(9,15) = **3** and LCM(9,15) = $3^2 \cdot 5 =$ **45**.

Note: If you have trouble seeing where the GCD and LCM are coming from here, it may help to insert the "missing" primes. In this case, 5 is missing from the factorization of 9. So it might help to write $9 = 3^2 \cdot 5^0$. Now we can think of the GCD as $3^1 \cdot 5^0 = 3$.

Solution 3:

9	15	**3**
3	5	3
1	5	5
1	1	

In the above table we started by placing the positive integers 9 and 15 in the upper left side of the table. We then started with the smallest prime 2. Since neither integer is divisible by 2, we move up to 3. Since 9 and 15 are both divisible by 3, we write 3 the right of 9 and 15, $\frac{9}{3} = 3$ below 9, and $\frac{15}{3} = 5$ below 15. We put the **3** in bold because *both* integers were divisible by 3.

Since 3 is divisible by 3, we write 3 to the right of 3 and 5, and $\frac{3}{3} = 1$ below 3. Note that we do not do anything to the 5 because 5 is not divisible by 3.

We repeat this process until we get two 1's on the left hand side. At this point we stop. The LCM is then the product of the primes in the right hand column. In this case LCM(9,15) = $3 \cdot 3 \cdot 5 = 3^2 \cdot 5 =$ **45**. The GCD is the product of the bold primes on the right. In this case GCD(9,15) = **3**.

Note: We can also use the fact that $\text{LCM}(a,b) \cdot \text{GCD}(a,b) = a \cdot b$ to get the GCD. We could use the equivalent form $\text{GCD}(a,b) = \frac{a \cdot b}{\text{LCM}(a,b)}$.

In this problem we have GCD(9,15) = $\frac{9 \cdot 15}{\text{LCM}(9,15)} = \frac{135}{45} = \mathbf{3}$.

Solution 4: On your TI-84 calculator press MATH, scroll right to NUM. For the GCD press 9, type 9, 15 and press ENTER. You will see an output of **3**. For the LCM press 8, type 9, 15 then ENTER for an output of **45**.

- If $\text{GCD}(a, b) = 1$, then a and b are said to be **relatively prime**.
- If a and b are consecutive positive integers, then a and b are relatively prime (i.e. $\text{GCD}(a, b) = 1$).
- If a and b are relatively prime, then $\text{LCM}(a, b) = a \cdot b$.
- If a, b, and c are three consecutive positive integers, then a, b, and c are relatively prime (that is $\text{GCD}(a, b, c) = 1$).
- $\text{LCM}(a, b) \cdot \text{GCD}(a, b) = a \cdot b$

Example 2.32: Find the GCD and LCM of 24 and 36.

Solution:

24	36	**2**
12	18	**2**
6	9	2
3	9	**3**
1	3	3
1	1	

LCM(24,36) = $2^3 \cdot 3^2 = \mathbf{72}$, and GCD(24,36) = $2^2 \cdot 3 = \mathbf{12}$.

Remarks: (1) We used the method of solution 3 from example 2.31 here. Note once again that if both numbers on the left change in the next step, then we put the number on the right in bold. The product of the bold numbers on the right gives the GCD, while the product of all the numbers on the right gives the LCM.

(2) We can also find the GCD by using $\text{LCM}(a, b) \cdot \text{GCD}(a, b) = a \cdot b$. In this case GCD(24,36) = $\frac{24 \cdot 36}{\text{LCM}(24,36)} = \frac{864}{72} = \mathbf{12}$.

Example 2.33: Find the GCD and LCM of 30 and 60.

Solution:

30	60	**2**
15	30	2
15	15	**3**
5	5	**5**
1	1	

LCM(30,60) = $2^2 \cdot 3 \cdot 5 = \mathbf{60}$, and GCD(30,60) = $2 \cdot 3 \cdot 5 = \mathbf{30}$.

Remark: Note that 30 is a divisor of 60.

In general, if a is a divisor of b, then GCD$(a, b) = a$ and LCM$(a, b) = b$.

Example 2.34: Find the GCD and LCM of 6, 8 and 24.

Solution:

6	8	24	**2**
3	4	12	2
3	2	6	2
3	1	3	3
1	1	1	

GCD(6,8,24) = 2 and LCM(6,8,24) = $2^3 \cdot 3 = \mathbf{24}$.

Remark: Note that 6 and 8 are both divisors of 24. So 24 is a multiple of 6, 8, and 24. Therefore LCM(6,8,24) = 24.

Example 2.35: Find the greatest common factor and least common multiple of 20, 44 and 72.

Solution:

20	44	72	**2**
10	22	36	**2**
5	11	18	2
5	11	9	3
5	11	3	3
5	11	1	5
1	11	1	11
1	1	1	

GCD(20,44,72) = $2^2 = \mathbf{4}$ and LCM(20,44,72) = $2^3 \cdot 3^2 \cdot 5 \cdot 11 = \mathbf{3960}$.

Example 2.36: Find GCD(5,12) and LCM(5,12).

Solution: Note that 5 and 12 are relatively prime. It follows that GCD(5,12) = **1** and LCM(5,12) = $5 \cdot 12 = \mathbf{60}$.

Remarks: (1) A quick way to see that 5 and 12 are relatively prime is to observe that 5 is prime, and 5 is *not* a divisor of 12.

(2) Recall that when two positive integers are relatively prime, their GCD is 1 and their LCM is the product of the two integers.

Example 2.37: Let a and b be positive integers such that $a = 12$, $\text{GCD}(a, b) = 3$, and $\text{LCM}(a, b) = 60$. What is b equal to?

Solution: Recall that $\text{LCM}(a, b) \cdot \text{GCD}(a, b) = a \cdot b$. So $60 \cdot 3 = 12b$. Therefore $12b = 180$, and so $b = \frac{180}{12} = \mathbf{15}$.

Example 2.38: Let a and b be positive integers such that $a \cdot b = 150$, and $\text{GCD}(a, b) = 5$. Find $\text{LCM}(a, b)$.

Solution: $\text{LCM}(a, b) \cdot \text{GCD}(a, b) = a \cdot b$. So $\text{LCM}(a, b) \cdot 5 = 150$. Therefore $\text{LCM}(a, b) = \frac{150}{5} = \mathbf{30}$.

Example 2.39: If $\text{GCD}(a, b) = 4$ and $\text{LCM}(a, b) = 112$, then $a \cdot b =$

Solution: $a \cdot b = \text{LCM}(a, b) \cdot \text{GCD}(a, b) = 4 \cdot 112 = \mathbf{448}$.

Example 2.40: A rectangular field whose width is 72 meters and whose length is 360 meters will be divided into equal squares to form several gardens. What is the least number of equal size square gardens that can be made so that there will be no field left?

Solution: We first find the GCD of 72 and 360.

72	360	**2**
36	180	**2**
18	90	**2**
9	45	**3**
3	15	**3**
1	5	5
1	1	

GCD(72,360) = $2^3 \cdot 3^2 = 72$.

So each square garden should have a length and width of 72

We now divide the area of the rectangular field by the area of a single square garden.

So the least number of square gardens that can be made is

$$\frac{\text{area of field}}{\text{area of garden}} = \frac{72 \cdot 360}{72 \cdot 72} = \mathbf{5}.$$

Example 2.41: 3 rolls of fabric containing 20 meters, 30 meters and 80 meters of fabric, respectively, will be cut into equal pieces. What is the least number of pieces of equal size fabric we can obtain so that each piece will have integer length and all the fabric is used?

Solution: We first find GCD(20,30,80).

20 30 80	**2**
10 15 40	2
5 15 20	2
5 15 10	2
5 15 5	3
5 5 5	**5**
1 1 1	

GCD(20,30,80) = $2 \cdot 5 = 10$.

$\frac{20}{10} = 2$, $\frac{30}{10} = 3$, and $\frac{80}{10} = 8$. So the least number of pieces of fabric we can obtain is 2 + 3 + 8 = **13**.

Example 2.42: Three pieces of cable which are 48 cm, 60 cm, and 90 cm long, are cut into equal pieces of integer length. What is the least number of pieces of cables that we can obtain in this manner?

Solution: We first find GCD(48,60,90).

48 60 90	**2**
24 30 45	2
12 15 45	2
6 15 45	2
3 15 45	**3**
1 5 15	3
1 5 5	5
1 1 1	

GCD(48, 60,90) = $2 \cdot 3 = 6$.

$\frac{48}{6} = 8$, $\frac{60}{6} = 10$, and $\frac{90}{6} = 15$. So the least number of pieces of cable of equal length we can obtain is 8 + 10 + 15 = **33**.

Example 2.43: When Ahmed stacks his books 4 at a time he has 3 books left over. When he stacks his books 5 at a time he again has 3 books left over. What is the least number of books that Ahmed can have?

Solution: We first find the LCM(4,5).

4 5	2
2 5	2
1 5	5
1 1	

LCM(4,5) = $2^2 \cdot 5 = 20$.

So the least number of books that Ahmed can have is 20 + 3 = **23**.

Example 2.44: Equal sized packages shaped like cubes with integer length sides, will be placed into a rectangular solid box with dimensions 50 cm, 60 cm, and 80 cm. What is the least number of such cube shaped packages that can be placed in the box so that there will be no empty space left inside the box.

Solution: We first find GCD(50,60,80).

50 60 80	**2**
25 30 40	2
25 15 20	2
25 15 10	2
25 15 5	3
25 5 5	**5**
5 1 1	5
1 1 1	

GCD(50,60,80) = $2 \cdot 5 = 10$.

So each cube should have a length, width, and height of 10 cm.

We now divide the volume of the box by the volume of a single cube.

So the least number of cubes that can be placed in the box with no space remaining is $\frac{\text{volume of box}}{\text{volume of cube}} = \frac{50\cdot60\cdot80}{10\cdot10\cdot10} = \mathbf{240}$.

Example 2.45: When Ahmed groups his pens 5 at a time, he has 2 pens left over. When he groups his pens 7 at a time, he again has 2 pens left over. What is the least number of pens Ahmed can have?

Solution: LCM(5,7) = 35. So the least number of pens that Ahmed can have is 35 + 2 = **37**.

Example 2.46: How many cube-shaped boxes with volume as large as possible can be placed into a box with edges of lengths 48 cm, 72 cm, and 96 cm, without leaving any empty space left over inside the box?

Solution: We first find GCD(48,72,96).

48	72	96	**2**
24	36	48	**2**
12	18	24	**2**
6	9	12	2
3	9	6	2
3	9	3	**3**
1	3	1	3
1	1	1	

GCD(48,72,96) = $2^3 \cdot 3 = 24$.

So each cube should have a length, width, and height of 24 cm.

We now divide the volume of the box by the volume of a single cube.

So the number of cubes that can be placed in the box with volume as large as possible and no space remaining is $\frac{\text{volume of box}}{\text{volume of cube}} = \frac{48 \cdot 72 \cdot 96}{24 \cdot 24 \cdot 24} = \mathbf{24.}$

Example 2.47: A carpenter wishes to cut equal size wooden cube pieces from a wooden rectangular solid with dimensions 12 cm, 20 cm, and 24 cm. How many equal size cube pieces with volume as large as possible can this carpenter obtain?

Solution: We first find GCD(12,20,24).

12	20	24	**2**
6	10	12	**2**
3	5	6	2
3	5	3	3
1	5	1	5
1	1	1	

GCD(12,20,24) = $2^2 = 4$.

So each cube should have a length, width, and height of 4 cm.

We now divide the volume of the rectangular solid by the volume of a single cube.

So the number of cubes that can be cut from the rectangular solid with volume as large as possible is $\frac{\text{volume of rectangular solid}}{\text{volume of cube}} = \frac{12 \cdot 20 \cdot 24}{4 \cdot 4 \cdot 4} = \mathbf{90}$.

Example 2.48: A can contains 63 ounces of olive oil and another can contains 84 ounces of sunflower oil. Both oils will be poured into bottles of equal volume so that there will be no oil remaining and each bottle is completely filled. What is the least number of bottles we need to do this?

Solution: GCD(63,84) = $3 \cdot 7 = 21$.

$\frac{63}{21} = 3$, and $\frac{84}{21} = 4$. So the least number of bottles we need is 3 + 4 = **7**.

3. RATIONAL NUMBERS

Definition: The set Q of rational numbers consists of all numbers of the form $\frac{a}{b}$ where a is an integer, and b is a nonzero integer.

Example 3.1: $\frac{-7}{3}$ and $\frac{8}{-5}$ are both rational numbers.

- Every natural number and every integer is a rational number.
- Rational numbers are also called *fractions*.
- Dividing zero by any nonzero integer results in zero.
- Division of any integer by zero is undefined.
- Each natural number and integer can be written as a fraction by writing a denominator of 1. For example -5 = $\frac{-5}{1}$ and 8 = $\frac{8}{1}$.
- The sign of rational numbers can be written in front of the integer in the numerator, denominator, or before the fraction bar.
- (+) signs can be omitted.

TYPES OF FRACTIONS

A) Proper Fractions: Without considering the sign, if the numerator of a fraction is less than the denominator, we call the fraction a *proper fraction*.

If $\frac{a}{b}$ is a proper fraction, then $-1 < \frac{a}{b} < 1$.

Example 3.2: $-\frac{3}{4}, -\frac{4}{5}, \frac{7}{9}$, and $\frac{10}{12}$ are all proper fractions.

B) Improper Fractions: Without considering the sign, if the numerator of a fraction is greater than or equal to the denominator, we call the fraction an *improper fraction*.

If $\frac{a}{b}$ is an improper fraction, then $\frac{a}{b} \leq -1$ or $\frac{a}{b} \geq 1$.

Example 3.3: $\frac{4}{3}, \frac{6}{3}, \frac{7}{2}$, and $\frac{10}{8}$ are all improper fractions.

C) Mixed Numbers: A mixed number is a number consisting of an integer and a proper fraction.

Example 3.4: $2\frac{1}{3}$, $7\frac{5}{6}$, and $-5\frac{2}{7}$ are all mixed numbers.

ADDITION OF FRACTIONS

A) Addition of Fractions with the Same Denominator:

$$\frac{a}{c} + \frac{b}{c} = \frac{a+b}{c}$$

Example 3.5: $\frac{7}{3} + \frac{6}{3} = \frac{13}{3}$, $\quad \frac{9}{5} + \frac{12}{5} = \frac{21}{5}$

B) Addition of Fractions with Different Denominators:

$$\frac{a}{c} + \frac{b}{d} = \frac{(a \cdot d + b \cdot c)}{c \cdot d}$$

Example 3.6: $\frac{4}{7} + \frac{5}{3} = \frac{4 \cdot 3 + 5 \cdot 7}{7 \cdot 3} = \frac{12+35}{21} = \frac{47}{21}$

Example 3.7: $\frac{3}{8} + \frac{4}{5} + \frac{9}{10} = \frac{3 \cdot 5 + 4 \cdot 8}{8 \cdot 5} + \frac{9}{10} = \frac{47}{40} + \frac{9}{10} = \frac{47 \cdot 10 + 9 \cdot 40}{40 \cdot 10} = \frac{830}{400}$

There is a more efficient way of adding fractions with different denominators by using the *least common multiple* of the denominators. Let's redo example 3.7 this way:

LCM(8, 5, 10) = $2^3 \cdot 5 = 40$.

$$\frac{3}{8} + \frac{4}{5} + \frac{9}{10} = \frac{3}{8} \cdot \frac{5}{5} + \frac{4}{5} \cdot \frac{8}{8} + \frac{9}{10} \cdot \frac{4}{4} = \frac{3 \cdot 5 + 4 \cdot 8 + 9 \cdot 4}{40} = \frac{15+32+36}{40} = \frac{83}{40}.$$

Notes: (1) We multiplied the numerator and denominator of each fraction by the integer necessary to change the denominator into the LCM. For example, Since $8 \cdot 5 = 40$, we multiplied the numerator and denominator of $\frac{3}{8}$ by 5.

(2) $\frac{830}{400} = \frac{83 \cdot 10}{40 \cdot 10} = \frac{83}{40} \cdot \frac{10}{10} = \frac{83}{40} \cdot 1 = \frac{83}{40}$.

(3) The number 10 used in Note 2 is the GCD (Greatest Common Divisor) of 830 and 400.

Example 3.8: $\frac{3}{7}+\frac{6}{14}+\frac{7}{3}=?$

Solution: LCM(7,14,3) = $2 \cdot 3 \cdot 7 = 42$.

$$\frac{3}{7}+\frac{6}{14}+\frac{7}{3}=\frac{3}{7}\cdot\frac{6}{6}+\frac{6}{14}\cdot\frac{3}{3}+\frac{7}{3}\cdot\frac{14}{14}=\frac{3\cdot 6+6\cdot 3+7\cdot 14}{42}=\frac{18+18+98}{42}=\frac{134}{42}$$

SUBTRACTION OF FRACTIONS

A) Subtraction of Fractions with the Same Denominator:

$$\frac{a}{c}-\frac{b}{c}=\frac{a-b}{c}$$

Example 3.9: $\frac{7}{10}-\frac{3}{10}=\frac{4}{10}$

B) Subtraction of Fractions with Different Denominators:

$$\frac{a}{c}-\frac{b}{d}=\frac{(a\cdot d-b\cdot c)}{c\cdot d}$$

Example 3.10: $\frac{7}{3}-\frac{6}{8}=\frac{7\cdot 8-6\cdot 3}{3\cdot 8}=\frac{56-18}{24}=\frac{38}{24}$

Note: In this particular example, the procedure for subtracting using the definition is identical to the procedure using the LCM. The reason for this is because 3 and 8 do not have any of the same prime factors.

MULTIPLICATION OF FRACTIONS

$$\frac{a}{c}\cdot\frac{b}{d}=\frac{a\cdot b}{c\cdot d}$$

Example 3.11: $\frac{2}{3}\cdot\frac{4}{7}=\frac{2\cdot 4}{3\cdot 7}=\frac{8}{21}$, $\frac{4}{3}\cdot\frac{5}{7}\cdot\frac{11}{13}=\frac{4\cdot 5\cdot 11}{3\cdot 7\cdot 13}=\frac{220}{273}$

DIVISION OF FRACTIONS

$$\frac{a}{c}\div\frac{b}{d}=\frac{a\cdot d}{c\cdot b}, b\neq 0$$

An easy way to remember how to divide two fractions is as follows: we keep the first fraction as it is, take the reciprocal of the second fraction, then multiply the two fractions.

$$\frac{a}{c}\div\frac{b}{d}=\frac{a}{c}\cdot\frac{d}{b}=\frac{a\cdot d}{c\cdot b}, b\neq 0$$

Example 3.12: $\frac{2}{3}\div\frac{11}{13}=\frac{2}{3}\cdot\frac{13}{11}=\frac{26}{33}$

Example 3.13: There are 60 students in a classroom. If $\frac{5}{12}$ of the students are girls, how many boys are there in the classroom?

Solution 1: There are $\frac{5}{12} \cdot 60 = \frac{300}{12} = 25$ girls in the classroom. Therefore the number of boys in the classroom is 60 – 25 = **35**.

Solution 2: Since $\frac{5}{12}$ of the students in the classroom are girls, it follows that $1 - \frac{5}{12} = \frac{12}{12} - \frac{5}{12} = \frac{12-5}{12} = \frac{7}{12}$ of the students are boys. So the number of boys in the classroom is $\frac{7}{12} \cdot 60 = \frac{420}{12} = \mathbf{35}$.

Solution 3:

Total number of students in the classroom = girls + boys

$$60 = \frac{5}{12} \cdot 60 + \text{boys}$$

$$60 = 25 + \text{boys}$$

So boys = 60 – 25 = **35**.

Remark: The word "of" can generally be translated as multiplication. For example "$\frac{5}{12}$ of 60" can be written as $\frac{5}{12} \cdot 60$.

Example 3.14: $\frac{1}{5}$ of the 800 students enrolled at a university went on a class trip. How many of the students did not go on the trip?

Solution: $\frac{4}{5} \cdot 800 = \mathbf{640}$.

Remark: This quick solution is just like Solution 2 from Example 3.13. You may want to try using the methods of Solutions 1 and 3 as well.

Example 3.15: Find $\frac{1}{3}$ of $\frac{1}{4}$ of 480

Solution 1: $\frac{1}{4}$ of 480 is $\frac{1}{4} \cdot 480 = 120$. So $\frac{1}{3}$ of $\frac{1}{4}$ of 480 is $\frac{1}{3}$ of 120, and $\frac{1}{3}$ of 120 is $\frac{1}{3} \cdot 120 = \mathbf{40}$.

Solution 2: "$\frac{1}{3}$ of $\frac{1}{4}$ of 480" = $\frac{1}{3} \cdot \frac{1}{4} \cdot 480 = \frac{480}{12} = \mathbf{40}$.

Remark: In order to find $\frac{1}{x}$ of $\frac{1}{y}$ of a number A, we multiply $\frac{1}{x} \cdot \frac{1}{y} \cdot A$.

Example 3.16: Ahmet travels $\frac{5}{9}$ of a 900 kilometer trip by car, $\frac{3}{4}$ of the rest of the trip by bus, and then walks for the remainder of the trip. How many kilometers did Ahmet walk?

Solution: Distance Ahmet traveled by car = $\frac{5}{9} \cdot 900 = 500$ km.

Distance Ahmet traveled by bus = $\frac{3}{4}(900 - 500) = \frac{3}{4} \cdot 400 = 300$ km.

Distance Ahmet walked = 900 – 500 – 300 = **100 km**.

Example 3.17: What is 20 more than $\frac{4}{9}$ of the number 144?

Solution: $\frac{4}{9}$ of 144 is $\frac{4}{9} \cdot 144 = 64$. So 20 more than $\frac{4}{9}$ of 144 is 20 more than 64. This is 64 + 20 = **84**.

Note: 20 more than 64 means to add 20 to 64. The correct way to write this is 64 + 20. Many students will write 20 + 64 instead. Although this computation gives the correct answer, it is technically incorrect. "20 + 64" is actually read "64 more than 20."

This particular mistake is pretty harmless, but if the question had the expression "less than" instead of "more than," then this mistake would actually lead to the wrong answer.

Quick solution: 20 more than $\frac{4}{9}$ of 144 is $\frac{4}{9} \cdot 144 + 20 = 64 + 20 = \mathbf{84}$.

Example 3.18: If x is a positive proper fraction, and $x + \frac{23}{50}$ is a natural number, find x.

Solution: $x = 1 - \frac{23}{50} = \frac{50}{50} - \frac{23}{50} = \frac{50-23}{50} = \mathbf{\frac{27}{50}}$.

Note: If $x + \frac{23}{50}$ were a natural number greater than 1, then x would have to be greater than 1, and therefore x would not be a *proper* fraction.

It follows that $x + \frac{23}{50} = 1$.

Example 3.19: If $\frac{3}{5} = \frac{x}{100}$, then what is the value of x ?

Solution 1: $\frac{3}{5} \cdot \frac{20}{20} = \frac{3 \cdot 20}{5 \cdot 20} = \frac{60}{100}$. So $x = \mathbf{60}$.

Solution 2: We multiply each side of the given equation by 100 to get

$$\frac{100}{1} \cdot \frac{3}{5} = \frac{x}{100} \cdot \frac{100}{1}$$

$$\frac{300}{5} = \frac{x}{1}$$

$$60 = x$$

So $x = \mathbf{60}$.

Solution 3: We cross multiply the given equation to get $5 \cdot x = 3 \cdot 100$. So $5x = 300$, and therefore $x = \frac{300}{5} = \mathbf{60}$.

Example 3.20: If $\frac{x}{9} = \frac{16}{5}$, then what is the value of x ?

Solution: $5x = 16 \cdot 9 = 144$. So $x = \mathbf{\frac{144}{5}}$.

Note: This solution is similar to Solution 3 from Example 3.19. You may want to try using the methods of Solutions 1 and 2 as well.

Example 3.21: $\frac{3}{4} + \left(-\frac{2}{7}\right) = ?$

Solution: LCM (4,7) = 28.

Therefore $\frac{3}{4} + \left(-\frac{2}{7}\right) = \frac{3}{4} \cdot \frac{7}{7} + \left(-\frac{2}{7}\right) \cdot \frac{4}{4} = \frac{21-8}{28} = \mathbf{\frac{13}{28}}$.

Remark: Since 4 and 7 have no common prime factors, the LCM is the same as the product $4 \cdot 7 = 28$.

Example 3.22: If $\frac{x+3}{4} = \frac{5}{12}$, then what is the value of x ?

Solution: $(x+3) \cdot 12 = 5 \cdot 4$. So $12x + 36 = 20$, and it follows that $12x = 20 - 36 = -16$. Finally, $x = -\frac{16}{12} = -\frac{4 \cdot 4}{3 \cdot 4} = -\frac{4}{3} \cdot \frac{4}{4} = \mathbf{-\frac{4}{3}}$.

Example 3.23: $\frac{5}{21} \cdot \frac{7}{30} = \frac{1 \cdot 35}{18 \cdot 35} = \mathbf{\frac{1}{18}}$.

COMPARING FRACTIONS

1) If two fractions have the same denominator, then the fraction with the greater numerator is greater than the other fraction.

If $a > b$, then $\frac{a}{c} > \frac{b}{c}$.

Note: The symbol ">" is read "is greater than." So $a > b$ means "a is greater than b."

Example 3.24: $\frac{7}{10} > \frac{6}{10} > \frac{5}{10}$

Example 3.25: $\frac{6}{\sqrt{3}} > \frac{5}{\sqrt{3}} > \frac{4}{\sqrt{3}}$

Example 3.26: $\frac{9}{\pi} > \frac{8}{\pi} > \frac{7}{\pi}$

Remark: The numbers given in examples 3.25 and 3.26 are *not* rational numbers. Nonetheless, the given inequalities still hold.

2) If two fractions have the same numerator, then the fraction with the greater denominator is smaller than the other fraction.

If $c > d$, then $\frac{a}{c} < \frac{a}{d}$.

Note: The symbol "$<$" is read "is less than." So $a < b$ means "a is less than b." Note that $a < b$ and $b > a$ have the same meaning. The symbol always points toward the smaller number.

Example 3.27: Arrange $\frac{13}{9}, \frac{13}{8}$, and $\frac{13}{7}$ in descending order.

Solution: Since these three fractions have the same numerator, we have $\frac{13}{9} < \frac{13}{8} < \frac{13}{7}$

Example 3.28: Arrange $\frac{\sqrt{17}}{5}, \frac{\sqrt{17}}{4}, \frac{\sqrt{17}}{6}$, and $\frac{\sqrt{17}}{3}$ in descending order.

Solution: Since the given fractions have the same numerator we have $\frac{\sqrt{17}}{3} > \frac{\sqrt{17}}{4} > \frac{\sqrt{17}}{5} > \frac{\sqrt{17}}{6}$.

Example 3.29: Arrange $\frac{\pi}{\sqrt{12}}, \frac{\pi}{\sqrt{15}}, \frac{\pi}{4}$, and $\frac{\pi}{\sqrt{17}}$ in descending order.

Solution: $\frac{\pi}{\sqrt{12}} > \frac{\pi}{\sqrt{15}} > \frac{\pi}{4} > \frac{\pi}{\sqrt{17}}$

3) Let $\frac{a}{b}$ and $\frac{c}{d}$ be positive fractions with $c \neq 0$.

- If $\frac{a}{b} \div \frac{c}{d} > 1$, then $\frac{a}{b} > \frac{d}{c}$

Example 3.30: $\frac{5}{7} \div \frac{2}{3} = \frac{5}{7} \cdot \frac{3}{2} = \frac{15}{14} > 1$. So $\frac{5}{7} > \frac{2}{3}$

- If $\frac{a}{b} \div \frac{c}{d} = 1$, then $\frac{a}{b} = \frac{d}{c}$
- If $\frac{a}{b} \div \frac{c}{d} < 1$, then $\frac{a}{b} < \frac{d}{c}$

Example 3.31: $\frac{4}{7} \div \frac{3}{5} = \frac{4}{7} \cdot \frac{5}{3} = \frac{20}{21} < 1$. So $\frac{4}{7} < \frac{3}{5}$

4) For positive fractions, if the difference between numerators and denominators are equal, then

→ for <u>proper</u> fractions, the value of the fraction <u>increases</u> as the numerators and denominators increase.

→ for <u>improper</u> fractions, the value of the fraction <u>decreases</u> as the numerators and denominators increase.

Example 3.32 : $\frac{6}{7} < \frac{9}{10} < \frac{15}{16}$

Note that the fractions are proper and (7 – 6) = (10 – 9) = (16 – 15) = 1

Example 3.33: $\frac{10}{12} < \frac{13}{15} < \frac{16}{18}$

Note that the fractions are proper and (12 – 10) = (15 – 13) = (18 – 16) = 2

Example 3.34: $\frac{11}{8} < \frac{9}{6} < \frac{7}{4}$

Note that the fractions are improper and (11 – 8) = (9 – 6) = (7 – 4) = 3

Example 3.35: $\frac{15}{11} < \frac{14}{10} < \frac{11}{7}$

Note that the fractions are improper and (15 – 11) = (14 – 10) = (11 – 7) = 4.

5) To compare negative fractions, first compare the corresponding positive fractions, and then reverse the resulting inequality.

Example 3.36: Compare $-\frac{4}{5}$ and $-\frac{3}{7}$.

Solution: We first compare $\frac{4}{5}$ and $\frac{3}{7}$. $\frac{4}{5} \div \frac{3}{7} = \frac{4}{5} \cdot \frac{7}{3} = \frac{28}{15} > 1$. So $\frac{4}{5} > \frac{3}{7}$.

It follows that $-\frac{4}{5} < -\frac{3}{7}$

Remark: Here are two other ways to compare $\frac{4}{5}$ and $\frac{3}{7}$:

Method 1: We change the denominators to LCM (5, 7) = 35

$\frac{4}{5} \cdot \frac{7}{7} = \frac{28}{35}$ and $\frac{3}{7} \cdot \frac{5}{5} = \frac{15}{35}$. Since the denominators are the same, the larger fraction has the *larger* denominator. So $\frac{28}{35} > \frac{15}{35}$ and so $\frac{4}{5} > \frac{3}{7}$.

Method 2: We change the numerators to LCM (3, 4) = 12

$\frac{4}{5} \cdot \frac{3}{3} = \frac{12}{15}$ and $\frac{3}{7} \cdot \frac{4}{4} = \frac{12}{28}$. Since the numerators are the same, the larger fraction has the *smaller* denominator. So $\frac{12}{15} > \frac{12}{28}$ and so $\frac{4}{5} > \frac{3}{7}$.

6) Let m and n be natural numbers. When the value of n is constant, the value of the proper fraction $\frac{m}{m+n}$ increases as the value of m increases.

Example 3.37: $\frac{1}{4} < \frac{2}{5} < \frac{5}{8} < \frac{11}{14}$

Note: In the previous example, $n = 3$, and m changes from 1 to 2 to 5 to 11, respectively.

Example 3.38: Write the fractions $\frac{5}{6}, \frac{11}{12}, \frac{7}{24},$ and $\frac{1}{4}$ in order from greatest to smallest.

Solution: We change the denominators to LCM (6, 12, 24, 4) = 24

$\frac{5}{6} \cdot \frac{4}{4} = \frac{20}{24}, \frac{11}{12} \cdot \frac{2}{2} = \frac{22}{24}, \frac{7}{24} \cdot \frac{1}{1} = \frac{7}{24},$ and $\frac{1}{4} \cdot \frac{6}{6} = \frac{6}{24}$

Since the denominators are the same, the larger fraction has the *larger* numerator. So $\frac{22}{24} > \frac{20}{24} > \frac{7}{24} > \frac{6}{24}$ and so $\frac{11}{12} > \frac{5}{6} > \frac{7}{24} > \frac{1}{4}$.

CHANGING MIXED NUMBERS TO IMPROPER FRACTIONS

$$a\frac{b}{c} = \frac{c \cdot a + b}{c}, \qquad -m\frac{n}{t} = -\left(\frac{t \cdot m + n}{t}\right)$$

Example 3.39: $2\frac{1}{3} = \frac{3 \cdot 2 + 1}{3} = \frac{7}{3}$

Example 3.40: $4\frac{3}{4} = \frac{4 \cdot 4 + 3}{4} = \frac{19}{4}$

Example 3.41: $-4\frac{5}{7} = -\left(\frac{7 \cdot 4 + 5}{7}\right) = -\frac{33}{7}$

Important note: $a\frac{b}{c} = a + \frac{b}{c}$ and $-m\frac{n}{t} = -m - \frac{n}{t}$.

Example 3.42: Evaluate $2004\frac{1}{3} - 2003\frac{1}{2}$

Solution 1: $2004\frac{1}{3} = \frac{3 \cdot 2004 + 1}{3} = \frac{6013}{3}$ and $2003\frac{1}{2} = \frac{2 \cdot 2003 + 1}{2} = \frac{4007}{2}$.

We change the denominators to LCM (2,3) = 6.

$\frac{6013}{3} \cdot \frac{2}{2} = \frac{12026}{6}, \frac{4007}{2} \cdot \frac{3}{3} = \frac{12021}{6}$.

So $2004\frac{1}{3} - 2003\frac{1}{2} = \frac{12026}{6} - \frac{12021}{6} = \mathbf{\frac{5}{6}}$.

Solution 2: $2004\frac{1}{3} = 2004 + \frac{1}{3}$ and $-2003\frac{1}{2} = -2003 - \frac{1}{2}$.

So $2004\frac{1}{3} - 2003\frac{1}{2} = 2004 + \frac{1}{3} - 2003 - \frac{1}{2}$

$= (2004 - 2003) + \left(\frac{1}{3} - \frac{1}{2}\right) = 1 + \frac{2-3}{6} = 1 - \frac{1}{6} = \frac{6}{6} - \frac{1}{6} = \frac{6-1}{6} = \mathbf{\frac{5}{6}}$.

Example 3.43: Evaluate $2\frac{1}{3} + 4\frac{2}{5}$

Solution: $2\frac{1}{3} + 4\frac{2}{5} = \frac{3 \cdot 2 + 1}{3} + \frac{5 \cdot 4 + 2}{5} = \frac{7}{3} + \frac{22}{5} = \frac{35 + 66}{15} = \mathbf{\frac{101}{15}}$.

EXPANDING AND REDUCING FRACTIONS

- If we multiply both the numerator and denominator of a fraction by the same nonzero number, the value of the fraction will not change.

$$\frac{a}{b}=\frac{a\cdot k}{b\cdot k}$$

- If we divide both the numerator and denominator of a fraction by the same nonzero number, the value of the fraction will not change.

$$\frac{a}{b}=\frac{a/k}{b/k}$$

Example 3.44: $\frac{2}{3}=\frac{2\cdot 4}{3\cdot 4}=\frac{8}{12}$

Note: In this example we **expanded** the fraction $\frac{2}{3}$ to $\frac{8}{12}$.

Example 3.45: $\frac{15}{24}=\frac{15/3}{24/3}=\frac{5}{8}$

Notes: (1) In this example we **reduced** the fraction $\frac{15}{24}$ to $\frac{5}{8}$.

(2) We say that $\frac{5}{8}$ is a **common fraction,** a **reduced fraction** or **in lowest terms** because 5 and 8 have no common factors except for 1. Another way to say this is that the **greatest common divisor** of 5 and 8 is 1 (or GCD(5,8) = 1)

(3) Another way to reduce $\frac{15}{24}$ is to first note that GCD (15, 24) = 3. We have $\frac{15}{24}=\frac{5\cdot 3}{8\cdot 3}=\frac{5}{8}\cdot\frac{3}{3}=\frac{5}{8}\cdot 1=\frac{5}{8}$.

Example 3.46: Expand the fraction $\frac{4}{7}$ by 9.

Solution: $\frac{4}{7}\cdot\frac{9}{9}=\frac{4\cdot 9}{7\cdot 9}=\mathbf{\frac{36}{63}}$.

Example 3.47: Expand $\frac{9}{11}$ by 5.

Solution: $\frac{9}{11}\cdot\frac{5}{5}=\frac{9\cdot 5}{11\cdot 5}=\mathbf{\frac{45}{55}}$.

DENSITY OF THE RATIONAL NUMBERS

The rational numbers are **dense**. This means that between any two rational numbers we can find another rational number.

One way to find a rational number between two given rational numbers is to first rewrite the two as fractions with the same denominator, then add the numerators, and place this new numerator over twice the common denominator.

* Let us find a fraction between $\frac{a}{b}$ and $\frac{c}{d}$.

$\frac{1}{2}\left(\frac{a}{b}+\frac{c}{d}\right)=\frac{\boldsymbol{ad+cb}}{\boldsymbol{2bd}}$

Note: All we are doing here is taking the average (arithmetic mean) of the two fractions.

Example 3.48: Find a fraction between $\frac{3}{4}$ and $\frac{7}{9}$.

Solution: $\frac{1}{2}\left(\frac{3}{4}+\frac{7}{9}\right)=\frac{3\cdot 9+7\cdot 4}{2\cdot 4\cdot 9}=\frac{27+28}{72}=\frac{\mathbf{55}}{\mathbf{72}}$.

Example 3.49: Find a fraction x such that $\frac{9}{7}<x<\frac{7}{5}$.

Solution: $\frac{1}{2}\left(\frac{9}{7}+\frac{7}{5}\right)=\frac{9\cdot 5+7\cdot 7}{2\cdot 7\cdot 5}=\frac{45+49}{70}=\frac{94}{70}=\frac{\mathbf{47}}{\mathbf{35}}$.

Example 3.50: Find a fraction between $\frac{1}{7}$ and $\frac{1}{6}$.

Solution: $\frac{1}{2}\left(\frac{1}{7}+\frac{1}{6}\right)=\frac{6+7}{2\cdot 7\cdot 6}=\frac{\mathbf{13}}{\mathbf{84}}$.

Example 3.51: Find a fraction between $\frac{2}{3}$ and $\frac{4}{7}$.

Solution: $\frac{1}{2}\left(\frac{2}{3}+\frac{4}{7}\right)=\frac{2\cdot 7+4\cdot 3}{2\cdot 3\cdot 7}=\frac{14+12}{42}=\frac{26}{42}=\frac{\mathbf{13}}{\mathbf{21}}$.

DECIMALS

A decimal is an integer followed by a decimal point and then finitely many or infinitely many digits. Some examples of decimals are

2.2 -0.235 1.2323232323… 3.14159265… 76.1010010001…

Fact: Every rational number can be written as a terminating or repeating decimal.

For example, $\frac{1}{2}=0.5$ and $\frac{1}{3}=0.3333333\ldots=0.\overline{3}$.

The line above the 3 indicates that the 3 keeps repeating.

The set of all decimals is the set R of **real numbers**. The set Q of **rational numbers** consists of the decimals which terminate or repeat. All other real numbers are called **irrational numbers** (sometimes written Q^c).

Some examples of irrational numbers are $\pi = 3.14159265\ldots$ and $76.1010010001\ldots$

Note that although the last number has a pattern, it is a nonrepeating pattern, and therefore it represents an irrational number.

REPEATING DECIMALS

A **repeating decimal** is a decimal in which a sequence of one or more digits is eventually repeated indefinitely.

- $a.bbb\ldots = a.\bar{b}$

For example, $0.777\ldots = 0.\bar{7}$ and $12.333\ldots = 12.\bar{3}$ are simple repeating decimals.

- $a.bcbcbc\ldots = a.\overline{bc}$

For example, $18.242424\ldots = 18.\overline{24}$ is a repeating decimal.

- $0.\bar{a} = \frac{a}{9}$

For example, $0.\bar{4} = \frac{4}{9}$, $0.\bar{5} = \frac{5}{9}$, and $0.\bar{6} = \frac{6}{9} = \frac{2}{3}$.

- $0.\overline{ab} = \frac{ab}{99}$

For example, $0.\overline{27} = \frac{27}{99} = \frac{3}{11}$, $0.\overline{33} = \frac{33}{99} = \frac{1}{3}$, and $0.\overline{11} = \frac{11}{99} = \frac{1}{9}$.

Note: $0.\overline{33}$ is just another way of writing $0.\bar{3}$ and similarly $0.\overline{11} = 0.\bar{1}$.

- $0.a\bar{b} = \frac{ab-a}{90}$

For example, $0.2\bar{3} = \frac{23-2}{90} = \frac{21}{90} = \frac{7}{30}$, and $0.1\bar{3} = \frac{13-1}{90} = \frac{12}{90} = \frac{2}{15}$.

- $a.b\bar{c} = a\frac{bc-b}{90}$

For example, $5.4\bar{3} = 5\frac{43-4}{90} = 5\frac{39}{90} = 5\frac{13}{30}$

- $a.bc\overline{de} = a\frac{bcde-bc}{9900}$

For example, $8.65\overline{43} = 8\frac{6543-65}{9900} = 8\frac{6478}{9900} = 8\frac{3239}{4950}$

Example 3.52: $0.\bar{3} + 0.\bar{4} = ?$

Solution: $0.\bar{3} + 0.\bar{4} = \frac{3}{9} + \frac{4}{9} = \frac{7}{9}$ or $\mathbf{0.\bar{7}}$.

Example 3.53: $0.\bar{6} - 0.\bar{4} = ?$

Solution: $0.\bar{6} - 0.\bar{4} = \frac{6}{9} - \frac{4}{9} = \frac{2}{9}$ or $\mathbf{0.\bar{2}}$.

Example 3.54: Express $3.\overline{4}$ as an improper fraction.

Solution: $3.\overline{4} = 3\frac{4}{9} = \frac{9\cdot3+4}{9} = \frac{27+4}{9} = \mathbf{\frac{31}{9}}$.

Example 3.55: $1.\overline{4} + 1.\overline{2} + 1.\overline{3} = ?$

Solution: $1.\overline{4} + 1.\overline{2} + 1.\overline{3} = 1\frac{4}{9} + 1\frac{2}{9} + 1\frac{3}{9} = \frac{9\cdot1+4}{9} + \frac{9\cdot1+2}{9} + \frac{9\cdot1+3}{9}$

$$= \frac{13}{9} + \frac{11}{9} + \frac{12}{9} = \frac{36}{9} = \mathbf{4}.$$

Example 3.56: $0.\overline{23} + 0.\overline{65} = ?$

Solution: $0.\overline{23} + 0.\overline{65} = \frac{23}{99} + \frac{65}{99} = \frac{88}{99} = \frac{8\cdot11}{9\cdot11} = \mathbf{\frac{8}{9}}$ or $.\overline{\mathbf{8}}$.

Example 3.57 : Express $0.3\overline{21}$ as a proper fraction.

Solution: $0.3\overline{21} = \frac{321-3}{990} = \frac{318}{990} = \mathbf{\frac{53}{165}}$.

Example 3.58: Express$4.3\overline{2}$ as an improper fraction.

Solution: $4.3\overline{2} = 4\frac{32-3}{90} = 4\frac{29}{90} = \frac{90\cdot4+29}{90} = \frac{360+29}{90} = \mathbf{\frac{389}{90}}$.

NOTE: If the repeating number is 9, then we can delete the repeating 9's and increase the preceding digit by 1.

For example, $5.\overline{9} = 6$, $4.3\overline{9} = 4.4$, $13.0\overline{9} = 13.1$, and $16.02\overline{9} = 16.03$.

Example 3.59: Express $3.1\overline{7}$ as an improper fraction.

Solution 1: $3.1\overline{7} = 3\frac{17-1}{90} = 3\frac{16}{90} = \mathbf{3\frac{8}{45}} = \frac{45\cdot3+8}{45} = \mathbf{\frac{143}{45}}$.

Solution 2: $3.1\overline{7} = \frac{317-31}{90} = \frac{286}{90} = \mathbf{\frac{143}{45}}$.

Example 3.60: $0.\overline{2} + 0.\overline{3} + 0.\overline{4} + 0.\overline{5} + 0.\overline{6} = ?$

Solution: $0.\overline{2} + 0.\overline{3} + 0.\overline{4} + 0.\overline{5} + 0.\overline{6} = \frac{2}{9} + \frac{3}{9} + \frac{4}{9} + \frac{5}{9} + \frac{6}{9} = \mathbf{\frac{20}{9}}$.

Example 3.61: If $A = 0.\overline{4} + 0.\overline{5}$ and $B = 0.\overline{3} + 0.\overline{2}$, then $\frac{A+B}{A-B} = ?$

Solution: $A = 0.\overline{4} + 0.\overline{5} = \frac{4}{9} + \frac{5}{9} = \frac{9}{9} = 1$, $B = 0.\overline{3} + 0.\overline{2} = \frac{3}{9} + \frac{2}{9} = \frac{5}{9}$.

So $\frac{A+B}{A-B} = \frac{1+\frac{5}{9}}{1-\frac{5}{9}} = \frac{\frac{14}{9}}{\frac{4}{9}} = \frac{14}{9}\cdot\frac{9}{4} = \frac{14}{4} = \mathbf{\frac{7}{2}}$.

ROUNDING DECIMALS

* To round a decimal to a given digit, we compare the first digit to the right of the given digit with 5:
    - If the digit to the right of the given digit is greater than or equal to 5, the given digit is increased by 1, digits to the right of this digit are omitted, and the new decimal is written accordingly.
    - If the digit to the right of the given digit is smaller than 5, the given digit remains unchanged, digits to the right of this digit are omitted, and the new decimal is written accordingly.

Example 3.62: Round 0.53 to the tenths digit.

Solution: The tenths digit is 5, and the digit to the right of tenths digit is 3. Since $3 < 5$, we do not add 1 to 5, and discard the 3. So, 0.53 rounded to the tenths digit is **0.5**.

Example 3.63: Round the decimal 0.378 to the hundredths digit.

Solution: The hundredths digit is 7, and the digit to the right of the hundredths digit is 8. Since $8 > 5$, we add 1 to 7, and discard all other numbers to the right. So, 0.378 rounded to the hundredths digit is **0.38**.

Example 3.64: Round the following numbers to the tenths digits.

5.444 5.2828 5.072

Solution: 5.4 5.3 5.1

Example 3.65: Round the following numbers to the hundredths digits.

6.378 6.3738

Solution: 6.38 6.37

Example 3.66: Round 17.54 to the nearest natural number.

Solution: The units digit is 7, and the digit to the right of the units digit is 5. Since $5 = 5$, we add 1 to 7, and discard all other numbers to the right. So, 17.54 rounded to the nearest natural number is **18**.

Example 3.67: Round 0.78 to the tenths digit.

Solution: The tenths digit is 7, and the digit to the right of the tenths digit is 8. Since $8 \geq 5$, we add 1 to 7, and discard all digits to the right. So 0.78 rounded to the tenths digit is **0.8**.

Example 3.68: Round 0.46 to the tenths digit.

Solution: The tenths digit is 4, and the digit to the right of the tenths digit is 6. Since $6 \geq 5$, we add 1 to 4, and discard all digits to the right. So 0.46 rounded to the tenths digit is **0.5**.

EXPANDED FORM OF A DECIMAL

- The integer part of a decimal consists of the digits to the left of the decimal point and the decimal part consists of the digits to the right of the decimal point.

Example 3.69: Write the decimal 346.435 in expanded form.

Solution: $3 \times 100 + 4 \times 10 + 6 \times 1 + 4 \times 0.1 + 3 \times 0.01 + 5 \times 0.001$

Notes: (1) As an alternative we can write the expanded form as

$$3 \times 100 + 4 \times 10 + 6 \times 1 + 4 \times \frac{1}{10} + 3 \times \frac{1}{100} + 5 \times \frac{1}{1000}.$$

(2) Another alternative:

$$3 \times 10^2 + 4 \times 10^1 + 6 \times 10^0 + 4 \times 10^{-1} + 3 \times 10^{-2} + 5 \times 10^{-3}.$$

Example 3.70: Write the decimal 64.508 in expanded form.

Solution: $6 \times 10 + 4 + 5 \times 0.1 + 0 \times 0.01 + 8 \times 0.001$

Notes: (1) As an alternative we can write the expanded form as

$$6 \times 10 + 4 \times 1 + 5 \times \frac{1}{10} + 0 \times \frac{1}{100} + 8 \times \frac{1}{1000}.$$

(2) Another alternative:

$$6 \times 10^1 + 4 \times 10^0 + 5 \times 10^{-1} + 0 \times 10^{-2} + 8 \times 10^{-2}.$$

4. SEQUENCES AND SERIES

ARITHMETIC SEQUENCES

Here is an example of an arithmetic sequence.

Example 4.1: 1, 4, 7, 10, 13, 16,...

Note that the first term of this sequence is 1, the second term of this sequence is 4, and so on.

So what makes this sequence arithmetic?

Well notice that to get from 1 to 4 we need to add 3. To get from 4 to 7 we also add 3. To get from 7 to 10 we also add 3. In other words, the sequence is arithmetic because we always add the same number to get from any term to the next term. This special number is called the **common difference** of the arithmetic sequence.

So why is this number called the common difference? Well another way to compute the common difference is to note that when we subtract any term from the next term we always get the same number, in this case that number is $d = 3$.

In other words we have $4 - 1 = 3$, $7 - 4 = 3$, and so on.

Definition: An **arithmetic sequence** is a sequence of numbers such that the difference $\boldsymbol{d}$ between consecutive terms is constant. The number $\boldsymbol{d}$ is called the **common difference** of the arithmetic sequence.

Example 4.2: The second term of an arithmetic sequence is 15 and the third term is 10. What is the first term?

Solution: Moving backwards, to get from the third term to the second term we add 5. Therefore we add 5 more to get to the first term. So the first term is 15 + 5 = **20**.

Notes: (1) In an arithmetic sequence, you always add (or subtract) the same number to get from one term to the next. This can be done by moving forwards or backwards through the sequence.

(2) The common difference of this sequence is $d = 10 - 15 = -5$.

Many students might mistakenly say that the common difference is 5.

(3) This particular problem was pretty simple, so we were able to solve it just by "counting." In other words we didn't really have to worry about the formalities of whether the common difference was positive or negative. But in harder questions we might need to be more careful.

Example 4.3: Find the common difference d of the arithmetic sequence 35, 31, 27, 23, 1,...

Solution: $d = 31 - 35 = \mathbf{-4}$.

ARITHMETIC SEQUENCES AND LINEAR EQUATIONS

Recall the formula for the slope of a line: Slope = $m = \frac{\text{rise}}{\text{run}} = \frac{y_2 - y_1}{x_2 - x_1}$

(Slope will be covered more formally in chapter 14.)

There is a natural correspondence between arithmetic sequences and linear equations. Each term of an arithmetic sequence can be naturally identified with a point on the corresponding line. Let's take the sequence from our first example: 1, 4, 7, 10, 13, 16,...

We can identify terms of the sequence with points on a line where the x-coordinate is the term number and the y-coordinate is the term itself.

$$(1,1), (2,4), (3,7), (4,10), (5,13), (6,16),\ldots$$

These points all lie on the same line, and we can compute the slope of this line by using any two of these points. For example, let's use the points (2,4) and (6,16).

We have $m = \frac{16-4}{6-2} = \frac{12}{4} = 3.$

Do you recognize this number? That's right! It's the common difference of the sequence.

The identification of arithmetic sequences with linear equations gives us a nice method for finding the common difference of an arithmetic sequence as long as we know **any** 2 terms of the sequence.

Example 4.4: Each term of a certain sequence is greater than the term before it. The difference between any two consecutive terms in the sequence is always the same number. If the fifth and ninth terms of the sequence are 33 and 97, respectively, what is the twelfth term?

Solution: We identify the two given terms with the points (5,33) and (9,97). The common difference is then $d = \frac{97-33}{9-5} = \frac{64}{4} = 16$. The twelfth term is then $97 + 16(3) = \mathbf{145}$.

Note that the ninth term is 97. So the tenth term is 97 + 16 = 113, the eleventh term is 113 + 16 = 129, and the twelfth term is 129 + 16 = 145.

ARITHMETIC SEQUENCE FORMULA

$$\textbf{Arithmetic sequence formula: } \boldsymbol{a_n = a_1 + (n-1)d}$$

In this formula, a_n is the nth term of the sequence. For example, a_1 is the first term of the sequence.

Example 4.5: In the arithmetic sequence 20, 15, 10, 5, 0, –5, –10,... we have that $a_1 = 20$ and $d = -5$. Therefore

$$a_n = 20 + (n-1)(-5) = 20 - 5n + 5 = \mathbf{25 - 5}\boldsymbol{n}.$$

So for example the 50th term of this arithmetic sequence is

$$a_{50} = 25 - 5(50) = -225.$$

Example 4.6: Find the general term of the arithmetic sequence whose first term is 8, and common difference is 3.

Solution: $a_n = a_1 + (n-1)d = 8 + (n-1)(3) = 8 + 3n - 3 = 5 + 3n$.

So $\boldsymbol{a_n = 3n + 5}$.

Example 4.7: Each term of a certain sequence is greater than the term before it. The difference between any two consecutive terms in the sequence is always the same number. If the fifth and ninth terms of the sequence are 33 and 97, respectively, what is the twelfth term?

Solution: Substituting 5 in for n and 33 in for a_n into the arithmetic sequence formula gives us

$$33 = a_1 + 4d$$

Similarly, substituting 9 in for n and 97 in for a_n into the arithmetic sequence formula gives us

$$97 = a_1 + 8d\ .$$

So we solve the following system of equations to find d.

$$\begin{aligned} 97 &= a_1 + 8d \\ \underline{33} &\underline{= a_1 + 4d} \\ 64 &= \qquad 4d \end{aligned}$$

The last equation comes from subtraction. We now divide each side of this last equation by 4 to get

$$d = 16.$$

Finally, we add 16 to 97 three times to get 97 + 16(3) = **145**.

Example 4.8: The first term of an arithmetic sequence is 10, and the common difference is 4. What is the 9th term of the sequence?

Solution: $a_n = 10 + (n-1) \cdot 4$. So $a_9 = 10 + 8 \cdot 4 = 10 + 32 = \mathbf{42}$.

Example 4.9: In the arithmetic sequence (a_n), $a_2 + a_6 = 20$. Find $a_3 + a_5$.

Solution: $a_3 + a_5 = a_2 + a_6 = \mathbf{20}$.

Note: In an arithmetic sequence (a_n), $\frac{a_{i+j}}{2} = \frac{a_i + a_j}{2}$.

For example, $a_4 = \frac{a_2 + a_6}{2}$ because the average of 2 and 6 is 4.

Also, $a_4 = \frac{a_3 + a_5}{2}$ because the average of 3 and 5 is 4.

It follows that $2a_4 = a_3 + a_5 = a_2 + a_6 = 20$.

Example 4.10: Consider the arithmetic sequence 13, 18, 23, 28, 33, 38,... For what value of n is the nth term of this sequence equal to 78?

Solution: $a_n = a_1 + (n-1)d = 13 + (n-1) \cdot 5 = 13 + 5n - 5 = 5n + 8$.

$a_n = 78 \Leftrightarrow 5n + 8 = 78 \Leftrightarrow 5n = 70 \Leftrightarrow n = \frac{70}{5} = \mathbf{14}$.

ARITHMETIC SERIES

An **arithmetic series** is the sum of the terms of an arithmetic sequence.

Example 4.11: The second term of an arithmetic sequence is 4 and the common difference is 3. What is the sum of the first 5 terms of this sequence?

Solution: The first term of the sequence is 4 – 3 = 1. So the first 5 terms of the sequence are 1, 4, 7, 10, 13.

It follows that the sum of the first 5 terms is 1 + 4 + 7 + 10 + 13 = **35**.

Example 4.12: What is the sum of the first 5 terms of the arithmetic sequence in which the 7th term is 10 and the 11th term is 16 ?

Solution: We can find the common difference of this arithmetic sequence with the computation

$$d = \frac{16-10}{11-7} = \frac{6}{4} = 1.5.$$

So the sixth term of the sequence is 10 – 1.5 = 8.5, the fifth term is 8.5 – 1.5 = 7, the fourth term is 7 – 1.5 = 5.5, the third term 5.5 – 1.5 = 4, the second term 4 – 1.5 = 2.5, and the first term 2.5 – 1.5 = 1.

So the sum of the first 5 terms is 1 + 2.5 + 4 + 5.5 + 7 = **20**.

ARITHMETIC SERIES FORMULA

$$A_n = n \cdot m$$

where n is the number of terms and m is the average (arithmetic mean) of the first and last term.

Example 4.13: As a very simple example, let's use the formula to add up the first 5 terms of the sequence 1, 4, 7, 10, 13, 16,...

Solution: The number of terms is $n = 5$, and the average of the first and fifth term is $m = \frac{1+13}{2} = 7$. So the sum of the first 5 terms of this arithmetic sequence is $5 \cdot 7 = \mathbf{35}$.

Example 4.14: What is the sum of the first 50 terms of the arithmetic sequence in which the 3rd term is 10 and the 15th term is 22?

Solution: The number of terms is $n = 50$. The common difference is $\frac{22-10}{15-3} = \frac{12}{12} = 1$. It follows that the first term is 8 and the fiftieth term is $8 + 49 \cdot 1 = 57$. So $m = \frac{8+57}{2} = 32.5$. So the sum is $50 \cdot 32.5 =$ **1625**.

Note: The arithmetic sequence formula can be used to find the fiftieth term.

Arithmetic sequence formula: $\boldsymbol{a_n = a_1 + (n-1)d}$

So $a_{50} = 8 + 49 \cdot 1 = 57$.

Example 4.15: $7 + 14 + 21 + \cdots + 112 = ?$

Solution: Here we have $n = \frac{112}{7} = 16$ and $m = \frac{7+112}{2} = \frac{119}{2}$. So

$$A_n = n \cdot m = 16 \cdot \frac{119}{2} = \mathbf{952}.$$

Remarks: (1) If the first term of an arithmetic sequence is equal to the common difference of the arithmetic sequence, we can find the number of terms n by simply dividing the last term by the common difference: $\frac{a_n}{d}$. This is how we got n in the solution above.

(2) More generally, the number of terms n can be found with the computation $n = \frac{a_n - a_1 + d}{d}$.

Example 4.16: $6 + 12 + 18 + \cdots + 120 = ?$

Solution: $n = \frac{120}{6} = 20$ and $m = \frac{6+120}{2} = \frac{126}{2} = 63$. So

$$A_n = n \cdot m = 20 \cdot 63 = \mathbf{1260}.$$

Example 4.17: $13 + 26 + 39 + \cdots + 130 = ?$

Solution: $n = \frac{130}{13} = 10$ and $m = \frac{13+130}{2} = \frac{143}{2}$. So

$$A_n = n \cdot m = 10 \cdot \frac{143}{2} = \mathbf{715}.$$

Example 4.18: $7 + 12 + 17 + \cdots + 82 = ?$

Solution: $n = \frac{82-7+5}{5} = 16$ and $m = \frac{7+82}{2} = \frac{89}{2}$. So

$$A_n = n \cdot m = 16 \cdot \frac{89}{2} = \mathbf{712}.$$

Remark: This time we used the formula $n = \frac{a_n - a_1 + d}{d}$.

GEOMETRIC SEQUENCES

Here is an example of a geometric sequence.

Example 4.19: 1, 2, 4, 8, 16, 32,…

Note that the first term of this sequence is 1, the second term of this sequence is 2, and so on.

So what makes this sequence geometric?

Well notice that to get from 1 to 2 we need to multiply by 2. To get from 2 to 4 we also multiply by 2. To get from 4 to 8 we also multiply by 2. In other words, the sequence is geometric because we always multiply by the same number to get from any term to the next term. This special number is called the **common ratio** of the geometric sequence.

So why is this number called the common ratio? Well another way to compute the common ratio is to note that when we divide any term by the previous term we always get the same number. In this case that number is $r = 2$.

In other words we have $2 \div 1 = 2$, $4 \div 2 = 2$, and so on.

Definition: A **geometric sequence** is a sequence of numbers such that the quotient $\boldsymbol{r}$ between consecutive terms is constant. The number $\boldsymbol{r}$ is called the **common ratio** of the geometric sequence.

Example 4.20: The second term of a geometric sequence is 27 and the third term is 81. What is the first term?

Solution: Moving backwards, to get from the third term to the second term we divide by 3. Therefore we divide 27 by 3 to get the first term. So the first term is $27 \div 3 = \mathbf{9}$.

Notes: (1) In a geometric sequence, you always multiply (or divide) by the same number to get from one term to the next. This can be done by moving forwards or backwards through the sequence.

(2) The common ratio of the sequence in this example is $r = 81 \div 27 = 3$.

(3) Do not accidentally divide "the wrong way" to get $\frac{1}{3}$.

(4) This particular problem was pretty simple, so we were able to solve it just by "counting." In other words we didn't really have to worry about the formalities of how to find the common ratio. But in harder questions we might need to be more careful.

NONCONSECUTIVE TERMS

If we know two consecutive terms of a geometric sequence, we can find the common ratio simply by dividing the second known term by the first known term. What if we know 2 terms of a geometric sequence, but they are not consecutive. In this case we can still find the common ratio quickly as follows. Suppose g_n and g_m are the nth and mth terms of a geometric sequence with $m > n$. Then the common ratio of the geometric sequence is

$$r = \sqrt[m-n]{\frac{g_m}{g_n}} = \left(\frac{g_m}{g_n}\right)^{\frac{1}{m-n}}$$

The above formula looks a lot more complicated than it is. In words it says, "divide the second known term by the first known term, and then take a root." Which root do we take? Well we take the distance between the two terms, and that's the root we use! Let's take the sequence from example 4.20:

$$1, 2, 4, 8, 16, 32, \ldots$$

Note that the second term of this sequence is 2 and the fifth term of this sequence is 16. Let's find the common ratio of the sequence using the second and fifth terms.

Step 1: Divide the fifth term by the second term: $16 \div 2 = 8$.

Step 2: Find the distance between the two terms: $5 - 2 = 3$.

Step 3: Take the appropriate root (in this case the cube root): $\sqrt[3]{8} = 2$.

So the common ratio is 2, as we already knew.

Note that if we plug the numbers into the formula above directly we get

$$r = \sqrt[5-2]{\frac{16}{2}} = \sqrt[3]{8} = 2.$$

Let's try another example.

Example 4.21: In a geometric sequence, the third term is 4 and the fifth term is 9. What is the seventh term in the sequence?

Solution: We need to find the common ratio of the sequence.

First we divide the fifth term of the sequence by the third term of the sequence to get $\frac{9}{4}$.

Next we note that the distance between the two terms is $5 - 3 = 2$.

The common ratio is therefore $\sqrt{\frac{9}{4}} = \frac{3}{2}$.

Since the fifth term is 9, we have that the sixth term is $9(\frac{3}{2}) = \frac{27}{2}$, and so the seventh term is $(\frac{27}{2})(\frac{3}{2}) = \mathbf{\frac{81}{4}}$.

Note: We can find the common ratio with a single computation:

$$r = \sqrt[5-3]{\frac{9}{4}} = \sqrt{\frac{9}{4}} = \frac{3}{2}.$$

GEOMETRIC SEQUENCE FORMULA

Geometric sequence formula: $g_n = g_1 \cdot r^{n-1}$

In this formula, g_n is the nth term of the sequence. For example, g_1 is the first term of the sequence.

Example 4.22: In the geometric sequence $1, \frac{1}{2}, \frac{1}{4}, \frac{1}{8}, \frac{1}{16}, \ldots$ we have $g_1 = 1$ and $r = \frac{1}{2}$. Therefore

$$g_n = 1 \cdot \left(\frac{1}{2}\right)^{n-1} = \left(\frac{1}{2}\right)^{n-1}.$$

So for example, the 10th term of this geometric sequence is

$$g_{10} = \left(\frac{1}{2}\right)^9 = \frac{1}{512}.$$

Example 4.23: Find the general term of the geometric sequence whose first five terms are 7, 14, 28, 56, 112,...

Solution: $g_1 = 7$ and $r = \frac{14}{7} = 2$. So $g_n = 7 \cdot 2^{n-1}$.

Example 4.23: The first and second terms of a geometric sequence are k and bk, in that order. What is the 500th term of the sequence in terms of b and k ?

Solution: The first term of the sequence is $g_1 = k$ and the common ratio is $r = \frac{bk}{k} = b$. So $g_n = kb^{n-1}$. Therefore $g_{500} = \boldsymbol{kb^{499}}$.

GEOMETRIC SERIES

A **geometric series** is the sum of the terms of a geometric sequence.

It turns out that computing an **infinite geometric series** is easier than computing a finite geometric series.

The sum G of an infinite geometric series with first term g and common ratio r with $-1 < r < 1$ is

$$G = \frac{g}{1-r}$$

If the common ratio r is greater than or equal to 1 or less than or equal to -1, then the geometric series has no sum.

Example 4.24: What is the sum of an infinite geometric series with common ratio $\frac{9}{10}$ whose first term is 10?

Solution: The sum is $\frac{10}{1-\frac{9}{10}} = \frac{10}{\frac{1}{10}} = 10 \div \frac{1}{10} = 10 \cdot 10 = \mathbf{100}$.

Example 4.25: If the sum of a given infinite geometric series is 160 and the common ratio is $\frac{1}{4}$. What is the third term of this series?

Solution: We are given that $r = \frac{1}{4}$, so that $\frac{g}{1-r} = \frac{g}{1-.25} = \frac{g}{.75}$. We are also given that $\frac{g}{1-r} = 160$ so that $\frac{g}{.75} = 160$. Multiplying each side of this equation by .75 gives $g = 160(.75) = 120$.

So the first term of the series is 120. To get each term after that we multiply by the common ratio. So the second term is $\frac{1}{4}(120) = 30$ and the third term is $\frac{1}{4}(30) = \mathbf{7.5}$.

Finite geometric series: The sum G_n of the first n terms of a geometric sequence with first term g and common ratio r is

$$G_n = \frac{g(1-r^n)}{1-r}$$

Example 4.26: Use the formula for a finite geometric series to add up the first 5 terms of the sequence 1, 2, 4, 8, 16, 32,…

Solution: The number of terms is $n = 5$, the first term is $g = 1$ and the common ratio is $r = 2$. So the sum of the first 5 terms of this geometric sequence is $\frac{1(1-2^5)}{1-2} = \frac{1-32}{-1} = -\frac{31}{-1} = \mathbf{31}$.

Example 4.27: Compute the sum of the first 8 terms of the following sequence: $1, -\frac{1}{2}, \frac{1}{4}, -\frac{1}{8}, \frac{1}{16}, -\frac{1}{32}, \ldots$

Solution: The first term of the sequence is $g = 1$ and the common ratio is $r = -\frac{1}{2}$. So the sum of the first 8 terms of this sequence is $\frac{1\left(1-(-\frac{1}{2})^8\right)}{1-(-\frac{1}{2})} = \mathbf{\frac{85}{128}}$.

ADDITIONAL SERIES FORMULAS

1) Sum of the first *n* consecutive positive integers

$$1+2+3+4+\cdots+n=\frac{n(n+1)}{2}$$

Example 4.28: $1+2+3+\cdots+25=?$

Solution: $1+2+3+\cdots+25=\frac{25(26)}{2}=\mathbf{325}$.

Example 4.29: $1+2+3+\cdots+100=?$

Solution: $1+2+3+\cdots+100=\frac{100(101)}{2}=\mathbf{5050}$.

2) Sum of the first *n* consecutive positive even integers

$$2+4+6+8+\cdots+2n=n(n+1)$$

Example 4.30: $2+4+6+\cdots+50=?$

Solution: : $2+4+6+\cdots+50=25(26)\ =\mathbf{650}$.

Note: $2n=50$, so that $n=\frac{50}{2}=25$.

Example 4.31: $2+4+6+\cdots+100=?$

Solution: : $2+4+6+\cdots+100=50(51)\ =\mathbf{2550}$.

Example 4.32: $2+4+6+\cdots+64=?$

Solution: : $2+4+6+\cdots+64=32(33)\ =\mathbf{1056}$.

3) Sum of the squares of the first *n* consecutive integers

$$1^2+2^2+3^2+\cdots+n^2=\frac{n(n+1)(2n+1)}{6}$$

Example 4.33: $1^2+2^2+3^2+\cdots+10^2=?$

Solution: $1^2+2^2+3^2+\cdots+10^2=\frac{10(11)(21)}{6}=\mathbf{385}$.

Example 4.34: $1^2+2^2+3^2+\cdots+15^2=?$

Solution: $1^2+2^2+3^2+\cdots+5=\frac{15(16)(31)}{6}=\mathbf{1240}$.

4) Sum of the cubes of the first *n* consecutive integers

$$1^3+2^3+3^3+\cdots+n^3=\left(\frac{n(n+1)}{2}\right)^2$$

Example 4.35: $1^3+2^3+3^3+\cdots+6^3=?$

Solution: $1^3+2^3+3^3+\cdots+6^3=\left(\frac{6(7)}{2}\right)^2=\mathbf{441}$.

Example 4.36: $1^3 + 2^3 + 3^3 + \cdots + 10^3 = ?$

Solution: $1^3 + 2^3 + 3^3 + \cdots + 10^3 = \left(\frac{10(11)}{2}\right)^2 = \mathbf{3025}$.

5) Sum of the first *n* consecutive positive odd integers

$$1 + 3 + 5 + 7 + \cdots + (2n - 1) = n^2$$

Example 4.37: $1 + 3 + 5 + \cdots + 21 = ?$

Solution: : $1 + 3 + 5 + \cdots + 21 = 11^2 = \mathbf{121}$.

Note: $2n - 1 = 21$, so that $2n = 21 + 1 = 22$ and so $n = \frac{22}{2} = 11$.

Example 4.38: $1 + 3 + 5 + \cdots + 99 = ?$

Solution: : $1 + 3 + 5 + \cdots + 99 = 50^2 = \mathbf{2500}$.

5. INEQUALITIES AND ABSOLUTE VALUE

INEQUALITIES

Inequalities are statements involving one of the symbols <, >, ≤, or ≥, read "less than," "greater than," "less than or equal to," and "greater than or equal to," respectively.

Inequality rules: Let x, y, a, and b be real numbers.

1. When the same number is added to or subtracted from both sides of an inequality, the inequality symbol remains unchanged.

$$x < y \Rightarrow x + a < y + a \qquad x < y \Rightarrow x - a < y - a$$

For example, $3 < 5 \Rightarrow 3 + 2 < 5 + 2 \Rightarrow 5 < 7$ and

$8 < 10 \Rightarrow 8 - 2 < 10 - 2 \Rightarrow 6 < 8$.

2. When both sides of an inequality are multiplied or divided by the same positive real number, the inequality symbol remains unchanged.

$$a > 0, x < y \Rightarrow x \cdot a < y \cdot a \qquad a > 0, x < y \Rightarrow \frac{x}{a} < \frac{y}{a}$$

For example, $3 < 5 \Rightarrow 3 \cdot 4 < 5 \cdot 4 \Rightarrow 12 < 20$ and

$8 < 10 \Rightarrow 8 \div 4 < 10 \div 4 \Rightarrow 2 < \frac{5}{2}$.

3. When both sides of an inequality are multiplied or divided by the same negative real number, the inequality symbol reverses.

$$a < 0, x < y \Rightarrow x \cdot a > y \cdot a \qquad a < 0, x < y \Rightarrow \frac{x}{a} > \frac{y}{a}$$

For example, $3 < 5 \Rightarrow 3(-4) > 5(-4) \Rightarrow -12 > -20$ and

$8 < 10 \Rightarrow 8 \div (-4) > 10 \div (-4) \Rightarrow -2 > -\frac{5}{2}$.

4. $x < y$ and $a < b \Rightarrow x + a < y + b$

For example, $3 < 4$ and $6 < 10 \Rightarrow 3 + 6 < 4 + 10 \Rightarrow 9 < 14$.

5. $x < y$ and $y < z \Rightarrow x < z$

6. $0 < a < b \Rightarrow 0 < a^n < b^n$ for any positive integer n

For example, $0 < 3 < 4 \Rightarrow 0 < 3^2 < 4^2 \Rightarrow 0 < 9 < 16$

7. $a < b < 0 \Rightarrow a^{2n-1} < b^{2n-1} < 0$ for any positive integer n

For example, $-4 < -3 < 0 \Rightarrow (-4)^3 < (-3)^3 < 0 \Rightarrow -64 < -27 < 0$

Note: If n is an integer, then $2n - 1$ is an odd integer.

$$n = 1 \Rightarrow 2n - 1 = 2(1) - 1 = 1$$
$$n = 2 \Rightarrow 2n - 1 = 2(2) - 1 = 3$$
$$n = 3 \Rightarrow 2n - 1 = 2(3) - 1 = 5$$
$$\cdots$$

8. $a < b < 0 \Rightarrow 0 < b^{2n} < a^{2n}$ for any positive integer n

For example, $-3 < -2 < 0 \Rightarrow 0 < (-2)^2 < (-3)^2 \Rightarrow 0 < 4 < 9$

9. $0 < x < 1 \Rightarrow 0 < x^n < x < 1$ for any positive integer n

10. $x \cdot y > 0 \Rightarrow (x < y \Leftrightarrow \frac{1}{x} > \frac{1}{y})$

Example 5.1: If $3 < x < 4$ and $2 < y < 6$, then $x + y$ must be between what two real numbers?

Solution:

$$\begin{array}{c} 3 < x < 4 \\ \underline{2 < y < 6} \\ \mathbf{5 < x + y < 10} \end{array}$$

Example 5.2: If $5 < x < 6$ and $3 < y < 7$, then $x - y$ must be between what two real numbers?

Solution: We first multiply each part of the inequality $3 < y < 7$ by -1 to get $-3 > -y > -7$, or equivalently $-7 < -y < -3$ (note that the inequalities reverse because we are multiplying by a negative number). Now we add:

$$\begin{array}{c} 5 < x < 6 \\ \underline{-7 < -y < -3} \\ \mathbf{-2 < x - y < 3} \end{array}$$

Example 5.3: If $3 < x < 4$ and $-2 < y < 4$, then $-3x + 4y$ must be between what two real numbers?

Solution: We first multiply each part of the inequality $3 < x < 4$ by -3 to get $-9 > -3x > -12$, or equivalently $-12 < -3x < -9$. Next, we multiply each part of the inequality $-2 < y < 4$ by 4 to get $-8 < 4y < 16$. Now we add:

$$\begin{array}{c} -12 < -3x < -9 \\ \underline{-8 < 4y < 16} \\ \mathbf{-20 < -3x + 4y < 7} \end{array}$$

INTERVALS

*** Closed Interval**

Inequality notation: $a \le x \le b$

Interval notation: $[a, b]$

Line graph:

a b

*** Open Interval**

Inequality notation: $a < x < b$

Interval notation: (a, b)

Line graph:

*** Half Open Intervals (or Half Closed Intervals)**

Inequality notation: $a < x \le b$

Interval notation: $(a, b]$

Line graph:

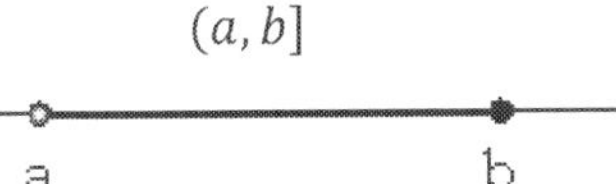

Inequality notation: $a \le x < b$

Interval notation: $[a, b)$

Line graph:

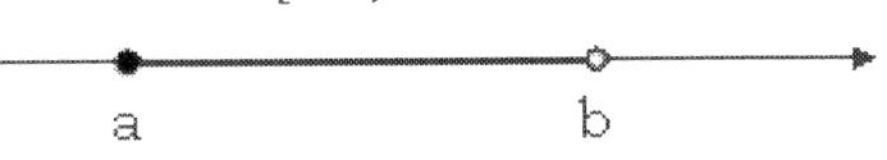

Example 5.4: Solve for x: $5x - 2 \ge 23$

Solution: $5x - 2 \ge 23 \Rightarrow 5x \ge 23 + 2 = 25 \Rightarrow x \ge \frac{25}{5} = 5$. So $\boldsymbol{x \ge 5}$.

Example 5.5: Solve for x: $11 \le 2x - 1 \le 19$

Solution: We add 1 to each side of the inequality to get $12 \le 2x \le 20$. Dividing by 2 gives $\mathbf{6 \le x \le 10}$, or in interval notation $[\mathbf{6}, \mathbf{10}]$. And here is the line graph:

Example 5.6: If $-2 < x \le 6$, find all possible values for x^2.

Solution: Since squaring a negative number gives the same value as squaring the corresponding positive number, we need only check $0 \le x \le 6$. So we have $0^2 \le x^2 \le 6^2$, or equivalently $\mathbf{0 \le x^2 \le 36}$.

Caution: A common mistake here is to square -2 to get $(-2)^2 = 4$ and write down the incorrect inequality $4 \le x^2 \le 36$.

Example 5.7: Solve for x: $6x + 4 \ge 10x - 20$

Solution: $6x + 4 \ge 10x - 20 \Rightarrow 6x - 10x \ge -20 - 4 \Rightarrow -4x \ge -24$.

We now divide each side of this last inequality by -4 to get $\mathbf{x \le 6}$.

Example 5.8: Solve for x: $6 - 3x < 0$

Solution: $6 - 3x < 0 \Rightarrow -3x < -6 \Rightarrow \mathbf{x > 2}$.

Example 5.9: Find the least integer value for which $6x - 24 \ge 0$.

Solution: $6x - 24 \ge 0 \Rightarrow 6x \ge 24 \Rightarrow x \ge 4$. So the least integer value is **4**.

DEFINITION OF ABSOLUTE VALUE

The **absolute value** of x, written $|x|$ is simply x if x is nonnegative, and $-x$ if x is negative. Put simply, $|x|$ just removes the minus sign if one is there.

Example 5.10: $|3| = 3$ $\quad |-3| = 3$ $\quad |0| = 0$ $\quad |\pi| = \pi$ $\quad |-3.5| = 3.5$

Geometrically, $|x - y|$ is the distance between x and y. In particular, $|x - y| = |y - x|$.

Example 5.11: $|5 - 3| = |3 - 5| = 2$ since the distance between 3 and 5 is 2.

If $|x - 3| = 7$, then the distance between x and 3 is 7. So there are two possible values for x. They are $3 + 7 = 10$, and $3 - 7 = -4$. See the figure below for clarification.

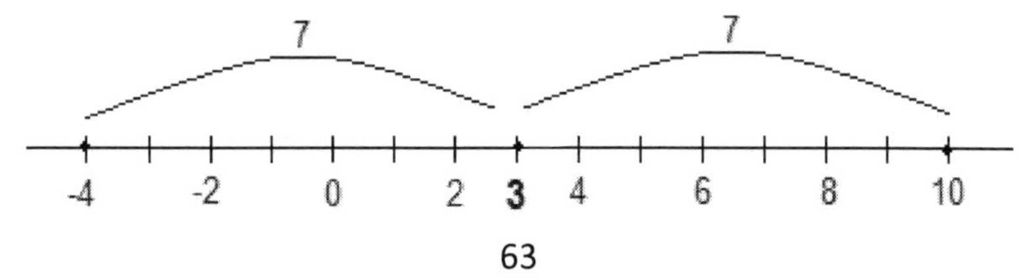

If $|x-3|<7$, then the distance between x and 3 is less than 7. If you look at the above figure you should be able to see that this is all x satisfying $-4<x<10$.

If $|x-3|>7$, then the distance between x and 3 is greater than 7. If you look at the above figure you should be able to see that this is all x satisfying $x<-4$ or $x>10$

Algebraically, we have the following. For $c>0$,

$$|x|=c \text{ is equivalent to } x=c \text{ or } x=-c$$

$$|x|<c \text{ is equivalent to } -c<x<c$$

$$|x|>c \text{ is equivalent to } x<-c \text{ or } x>c.$$

Let's look at the same examples as before algebraically.

Example 5.12: If $|x-3|=7$, then $x-3=7$ or $x-3=-7$. So $x=10$ or $x=-4$.

If $|x-3|<7$, then $-7<x-3<7$. So $-4<x<10$.

If $|x-3|>7$, then $x-3<-7$ or $x-3>7$. So $x<-4$ or $x>10$.

PROPERTIES OF ABSOLUTE VALUE

1) $|x|=|-x|$
2) $|x-y|=|y-x|$
3) $|x|\cdot|y|=|x\cdot y|$
4) $\frac{|x|}{|y|}=\left|\frac{x}{y}\right|$
5) If $x^2<y^2$, then $|x|<|y|$
6) $|x^n|=|x|^n$
7) If $|a|<|b|$, then $-|b|<|a|<|b|$
8) If $|x\cdot y|=0$, then $x=0$ or $y=0$
9) If $|x|+|y|=0$, then $x=y=0$
10) If $|x|=a$, then $x=a$ or $x=-a$

Example 5.13: What is the product of the solutions of the equation $|x-6|=4$?

Solution: By property 10 we have $x-6=4$ or $x-6=-4$. So $x=4+6=10$ or $x=-4+6=2$. So the equation $|x-6|=4$ has the two solutions $x=10$ and $x=2$. The product of these solutions is $10\cdot 2=\mathbf{20}$.

Example 5.14: The sum of the solutions of $|x-4|+|4-x|=8$ is

Solution: By property 2 we have $|4-x|=|x-4|$. So we have that $|x-4|+|4-x|=|x-4|+|x-4|=2|x-4|$. Therefore we have $2|x-4|=8$, and so $|x-4|=\frac{8}{2}=4$.

By property 10 we have $x - 4 = 4$ or $x - 4 = -4$. So $x = 4 + 4 = 8$ or $x = -4 + 4 = 0$. So the equation $|x - 6| = 4$ has the two solutions $x = 8$ and $x = 0$. The sum of these solutions is $8 + 0 = \mathbf{8}$.

Example 5.15: Evaluate the sum of the solutions of the equation $|x - 2| = |3x - 4|$.

Solution: We square each side of the given equation, and solve the resulting quadratic equation for x.

$$|x - 2|^2 = |3x - 4|^2$$
$$x^2 - 4x + 4 = 9x^2 - 24x + 16$$
$$0 = 8x^2 - 20x + 12$$

Dividing each side of this last equation by 4 gives $2x^2 - 5x + 3 = 0$. So $(2x - 3)(x - 1) = 0$, and $2x - 3 = 0$ or $x - 1 = 0$. So $x = \frac{3}{2}$ and $x = 1$ are solutions of the quadratic equation. We need to check to see if they are solutions of the original absolute value equation.

$x = \frac{3}{2}$: $|x - 2| = \left|\frac{3}{2} - 2\right| = \left|-\frac{1}{2}\right| = \frac{1}{2}$ and $|3x - 4| = \left|\frac{9}{2} - 4\right| = \left|\frac{1}{2}\right| = \frac{1}{2}$.

$x = 1$: $|x - 2| = |1 - 2| = |-1| = 1$ and $|3x - 4| = |3 - 4| = |-1| = 1$.

So $x = \frac{3}{2}$ and $x = 1$ are both solutions of the given equation. So the sum of the solutions is $\frac{3}{2} + 1 = \frac{5}{2}$ or $\mathbf{2.5}$.

Example 5.16: If $a > 0$, then $|4a| + |-3a| =$

Solution: Since $a > 0$, $|4a| + |-3a| = |4| \cdot |a| + |-3| \cdot |a| = 4a + 3a = \mathbf{7a}$.

Example 5.17: If $a < 0 < b < z$, then $\frac{|2a|}{2a} + \frac{|3b|}{3b} + \frac{4z}{4z} =$

Solution: $\frac{|2a|}{2a} + \frac{|3b|}{3b} + \frac{4z}{4z} = \frac{-2a}{2a} + \frac{3b}{3b} + \frac{4z}{4z} = -1 + 1 + 1 = \mathbf{1}$.

Notes: (1) Since $a < 0$, $|a| = -a$.

(2) Since $b > 0$, $|b| = b$.

(3) The sign of z is irrelevant. $\frac{4z}{4z} = 1$ as long as $z \neq 0$.

Example 5.18: If $|x + 6| + |6x + 2y| = 0$, then $\frac{2y-x}{2x+y} =$

Solution: By property 9, $x + 6 = 0$ and $6x + 2y = 0$. From the first equation, $x = -6$, and so the second equation gives $6(-6) + 2y = 0$, or equivalently $-36 + 2y = 0$. So $2y = 36$, and therefore $y = \frac{36}{2} = 18$.

So we have $\frac{2y-x}{2x+y} = \frac{2(18)-(-6)}{2(-6)+18} = \frac{36+6}{-12+18} = \frac{42}{6} = \mathbf{7}$.

Example 5.19: Find all x-values such that $|x - 4| = 6x - 20$.

Solution: If $x \geq 4$, then $x - 4 \geq 0$, and so $|x - 4| = x - 4$. Therefore, in this case we have $x - 4 = 6x - 20$, and so $5x = 16$. Therefore $x = \frac{16}{5}$. Since $\frac{16}{5} \ngeq 4$, $x = \frac{16}{5}$ is *not* a solution.

If $x < 4$, then $x - 4 < 0$, and so $|x - 4| = -(x - 4) = -x + 4 = 4 - x$. Therefore, in this case we have $4 - x = 6x - 20$, and so $7x = 24$. Therefore $x = \frac{24}{7}$. Since $\frac{24}{7} < 4$, $x = \frac{24}{7}$ is a solution.

So the only solution is $x = \mathbf{\frac{24}{7}}$.

Example 5.20: If $x - 3 = y$, then $|2x - 2y| + |3y - 3x| =$

Solution: Since $x - 3 = y$, we have $x - y = 3$. So

$$|2x - 2y| + |3y - 3x| = |2(x - y)| + |3(y - x)|$$
$$= |2| \cdot |x - y| + |3| \cdot |y - x| = 2|x - y| + 3|x - y| = 5|x - y|$$
$$= 5 \cdot 3 = \mathbf{15}.$$

Example 5.21: If $|a - 4| + |b - 6| + |c - 8| = 0$, then $\frac{a+b}{c} =$

Solution: By property 9, $a - 4 = 0$, $b - 6 = 0$, and $c - 8 = 0$. Therefore $a = 4$, $b = 6$, and $c = 8$. So $\frac{a+b}{c} = \frac{4+6}{8} = \frac{10}{8} = \mathbf{\frac{5}{4}}$.

Example 5.22: How many integers satisfy the inequality $|x + 4| \leq 8$?

Solution: $|x + 4| \leq 8$ is equivalent to $-8 \leq x + 4 \leq 8$. Subtracting 4 gives $-12 \leq x \leq 4$.

So the number of *integer* solutions is $4 - (-12) + 1 = 4 + 12 + 1 = \mathbf{17}$.

Notes: (1) The number of integers from a to b, inclusive, is $b - a + 1$.

For example, let's count the number of integers from 5 to 12, inclusive. They are 5, 6, 7, 8, 9, 10, 11, 12, and we see that there are 8 of them. Now 12 – 5 = 7 which is not the correct amount, but 12 – 5 + 1 = 8 which is the correct amount.

If you ever happen to forget this little formula test it out on a small list of numbers as I just did. But it's nice to have this one committed to memory so that it is there for you when you need it.

(2) Here is a graph illustrating the integer solutions:

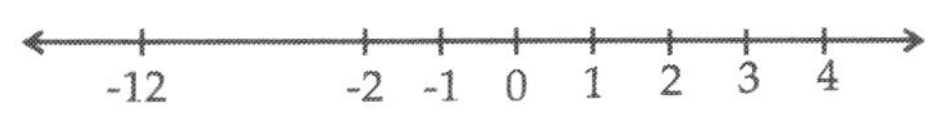

Example 5.23: $|\sqrt{8} - \pi| =$

Solution: $\pi \approx 3.14$, and $\sqrt{8} < 3$. Therefore $\pi > \sqrt{8}$. It follows that $|\sqrt{8} - \pi| = -(\sqrt{8} - \pi) = \boldsymbol{\pi - \sqrt{8}}$.

Example 5.24: If $a < 0 < b$, then $|a| + |b| + |2a - 2| =$

Solution: Since $a < 0$, $|a| = -a$. Also, since $a < 0$, we have $a - 1 < 0$, and so $|a - 1| = -(a - 1) = -a + 1 = 1 - a$. Since $b > 0$, $|b| = b$.

So $|a| + |b| + |2a - 2| = |a| + |b| + |2| \cdot |a - 1| = -a + b + 2(1 - a)$

$$= -a + b + 2 - 2a = \boldsymbol{-3a + b + 2}.$$

Example 5.25: If $3 < x < 4$, then $|x - 3| + |x - 4| =$

Solution: Since $x > 3$, $x - 3 > 0$, and so $|x - 3| = x - 3$.

Since $x < 4$, $x - 4 < 0$, and so $|x - 4| = -(x - 4) = -x + 4 = 4 - x$.

So $|x - 3| + |x - 4| = (x - 3) + (4 - x) = \mathbf{1}$.

Example 5.26: If $x < 0$, then $\frac{|-3x|}{x} + \frac{|4x|}{x} =$

Solution:

$$\frac{|-3x|}{x} + \frac{|4x|}{x} = \frac{|-3| \cdot |x|}{x} + \frac{|4| \cdot |x|}{x} = \frac{3(-x)}{x} + \frac{4(-x)}{x} = \frac{-3x}{x} + \frac{-4x}{x} = -3 - 4 = \mathbf{-7}.$$

Example 5.27: If $0 < a < b$, then $|2a - 2b| + |2a| + |2b| =$

Solution: Since $a, b > 0$, we have $|a| = a$ and $|b| = b$. Since $a < b$, we have $b - a > 0$, and so $|a - b| = |b - a| = b - a$.

So $|2a - 2b| + |2a| + |2b| = |2(a - b)| + |2| \cdot |a| + |2| \cdot |b|$

$= |2| \cdot |a - b| + 2a + 2b = 2(b - a) + 2a + 2b = 2b - 2a + 2a + 2b = \boldsymbol{4b}$.

Example 5.28: Find the solution set of $|x + 5| \le 6$.

Solution: $|x + 5| \le 6 \Rightarrow -6 \le x + 5 \le 6 \Rightarrow \boldsymbol{-11 < x < 1}$.

6. THE THREE MEANS

ARITHMETIC MEAN

The **arithmetic mean** (or **average)** of a list of numbers is the sum of the numbers in the list divided by the quantity of the numbers in the list.

$$\textbf{Average} = \frac{\textbf{Sum}}{\textbf{Number}}$$

- The arithmetic mean of a and b is $\frac{a+b}{2}$.
- The arithmetic mean of a, b, and c is $\frac{a+b+c}{3}$.
- The arithmetic mean of a, b, c and d is $\frac{a+b+c+d}{4}$.

Example 6.1: Find arithmetic mean of 64 and 84.

Solution: $\frac{64+84}{2} = \frac{148}{2} = \mathbf{74}$.

Example 6.2: Calculate the arithmetic mean of 30, 40 and 50.

Solution: $\frac{30+40+50}{3} = \frac{120}{3} = \mathbf{40}$.

Example 6.3: Compute the average of $\sqrt{12}$, $\sqrt{27}$, and $\sqrt{75}$.

Solution: $\frac{\sqrt{12}+\sqrt{27}+\sqrt{75}}{3} = \frac{2\sqrt{3}+3\sqrt{3}+5\sqrt{3}}{3} = \mathbf{\frac{10\sqrt{3}}{3}}$.

Notes: (1) $\sqrt{12} = \sqrt{4 \cdot 3} = \sqrt{4} \cdot \sqrt{3} = 2\sqrt{3}$.

(2) $\sqrt{27} = \sqrt{9 \cdot 3} = \sqrt{9} \cdot \sqrt{3} = 3\sqrt{3}$.

(3) $\sqrt{75} = \sqrt{25 \cdot 3} = \sqrt{25} \cdot \sqrt{3} = 5\sqrt{3}$.

Example 6.4: The heights of six students are 60, 64, 80, 88, 90 and 94 inches, respectively. What is the average height, in inches, of these six students?

Solution: $\frac{60+64+80+88+90+94}{6} = \frac{476}{6} = \mathbf{79.\overline{3}}$.

Example 6.5: Evaluate the arithmetic mean of the first ten positive even integers.

Solution: $\frac{2+4+6+8+10+12+14+16+18+20}{10} = \frac{110}{10} = \mathbf{11}$.

Example 6.6: Calculate the arithmetic mean of –10, 12, 20 and 30.

Solution: $\frac{-10+12+20+30}{4} = \frac{52}{4} = \mathbf{13}$.

Example 6.7: Calculate the average of the first ten prime numbers.

Solution: $\frac{2+3+5+7+11+13+17+19+23+29}{10} = \frac{129}{10} = \mathbf{12.9}$.

GEOMETRIC MEAN

The **geometric mean** (also **geometric average**) of a list of n numbers is the nth root of the product of the numbers in the list.

$$\textbf{Geometric Mean} = \sqrt[n]{\boldsymbol{a_1 \cdot a_2 \cdot a_3 \cdots a_n}}$$

* The geometric mean of a and b is $\sqrt{ab}$.
* The geometric mean of a, b, and c is $\sqrt[3]{abc}$.
* The geometric mean of a, b, c and d is $\sqrt[4]{abcd}$.

Example 6.8: Calculate the geometric mean of 25 and 64.

Solution: $\sqrt{25 \cdot 64} = \sqrt{25}\sqrt{64} = 5 \cdot 8 = \mathbf{40}$.

Example 6.9: Evaluate the geometric mean of 16 and 100.

Solution: $\sqrt{16 \cdot 100} = \sqrt{16}\sqrt{100} = 4 \cdot 10 = \mathbf{40}$.

Example 6.10: Evaluate the geometric mean of x^2 and y^2.

Solution: $\sqrt{x^2y^2} = \sqrt{x^2}\sqrt{y^2} = |x||y| = |xy|$.

Example 6.11: Evaluate the geometric mean of 2, 4 and 8.

Solution: $\sqrt[3]{2 \cdot 4 \cdot 8} = \sqrt[3]{2^1 2^2 2^3} = \sqrt[3]{2^6} = 2^{\frac{6}{3}} = 2^2 = \mathbf{4}$.

Example 6.12: Suppose that the geometric mean of 9 and x is 12. Find the arithmetic mean of 9 and x.

Solution: We are given that $\sqrt{9x} = 12$, so that $9x = 12^2 = 144$. Therefore $x = \frac{144}{9} = 16$. So the arithmetic mean of 9 and x is $\frac{9+x}{2} = \frac{9+16}{2} = \mathbf{\frac{25}{2}}$ or **12.5**.

Example 6.13: Find the geometric mean of the first three prime numbers.

Solution: The first three prime numbers are 2, 3 and 5. So the geometric mean is $\sqrt[3]{2 \cdot 3 \cdot 5} = \sqrt[3]{\mathbf{30}}$.

Example 6.14: Find the geometric mean of the prime numbers less than 15 and greater 5.

Solution: The prime numbers between 5 and 15 are 7, 11 and 13. So the geometric mean is $\sqrt[3]{7 \cdot 11 \cdot 13} = \sqrt[3]{\mathbf{1001}}$.

Example 6.15: Evaluate the ratio of the arithmetic mean of the first four positive even integers to the geometric mean of these integers.

Solution: The first four positive even integers are 2, 4, 6 and 8.

Arithmetic mean = $\frac{2+4+6+8}{4} = \frac{20}{4} = 5$.

Geometric mean =

$$\sqrt[4]{2 \cdot 4 \cdot 6 \cdot 8} = \sqrt[4]{2^1 \cdot 2^2 \cdot 2^1 \cdot 3 \cdot 2^3} = \sqrt[4]{2^7 \cdot 3} = \sqrt[4]{2^4}\sqrt[4]{2^3 \cdot 3} = \mathbf{2\sqrt[4]{24}}.$$

$$\frac{\text{Arithmetic mean}}{\text{Geometric mean}} = \mathbf{\frac{5}{2\sqrt[4]{24}}}$$

HARMONIC MEAN

The **harmonic mean** of a list of n numbers is the quantity of numbers in the list divided by the sum of the reciprocals of the numbers in the list.

$$\textbf{Harmonic Mean} = \frac{n}{\frac{1}{a_1}+\frac{1}{a_2}+\frac{1}{a_3}+\cdots+\frac{1}{a_n}}$$

- The harmonic mean of a and b is $\frac{2}{\frac{1}{a}+\frac{1}{b}}$.
- The harmonic mean of a, b, and c is $\frac{3}{\frac{1}{a}+\frac{1}{b}+\frac{1}{c}}$.
- The harmonic mean of a, b, c and d is $\frac{4}{\frac{1}{a}+\frac{1}{b}+\frac{1}{c}+\frac{1}{d}}$.

Example 6.16: Find harmonic mean of 3 and 4.

Solution: $\frac{2}{\frac{1}{3}+\frac{1}{4}} = \frac{2}{\frac{4+3}{3\cdot 4}} = 2 \div \frac{7}{12} = 2 \cdot \frac{12}{7} = \mathbf{\frac{24}{7}}$.

Example 6.17: Find the harmonic mean of $\frac{1}{3}$, 6 and 10.

Solution: $\frac{3}{3+\frac{1}{6}+\frac{1}{10}} = \frac{3}{\frac{90+5+3}{30}} = 3 \div \frac{98}{30} = 3 \cdot \frac{30}{98} = \frac{90}{98} = \mathbf{\frac{45}{49}}$.

Note: LCM(6,10) = 30

Example 6.18: Find the harmonic mean of 3, 4 and 5.

Solution: $\frac{3}{\frac{1}{3}+\frac{1}{4}+\frac{1}{5}} = \frac{3}{\frac{20+15+12}{60}} = 3 \div \frac{47}{60} = 3 \cdot \frac{60}{47} = \mathbf{\frac{180}{47}}$.

Note: LCM(3,4,5) = 60

Example 6.19: Find the harmonic mean of 2, 4, 6 and 8.

Solution: $\frac{4}{\frac{1}{2}+\frac{1}{4}+\frac{1}{6}+\frac{1}{8}} = \frac{4}{\frac{12+6+4+3}{24}} = 4 \div \frac{25}{24} = 4 \cdot \frac{24}{25} = \mathbf{\frac{96}{25}}$.

7. PROPORTIONALITY

RATIOS

A **ratio** is the quotient of two quantities (at least one non-zero) of the same kind (same units).

Example 7.1: The ratio of 15 kg sugar to 3 kg salt is $\frac{15\text{ kg}}{3\text{ kg}} = \mathbf{5}$.

Example 7.2: The ratio of 25 gal of water to 45 gal of milk is $\frac{25\text{ gal}}{45\text{ gal}} = \mathbf{\frac{5}{9}}$.

Example 7.3: Calculate the ratio of milk to the total mixture when 25 gal of water is mixed with 45 gal of milk.

Solution: $\frac{45}{25+45} = \frac{45}{70} = \mathbf{\frac{9}{14}}$.

A **proportion** is an equation stating that two ratios are equal: $\frac{a}{b} = \frac{c}{d}$

Notes: (1) $\frac{a}{b} = \frac{c}{d}$ can also be written $a: b = c: d$, or $ad = bc$.

(2) $\frac{a}{b} = \frac{c}{d} = k$ is sometimes called a **dual proportion**. It can also be expressed as $a: c = b: d$ or $a: b = c: d$. It follows that $ad = bc$.

(3) $\frac{a}{b} = \frac{c}{d} = \frac{e}{f} = k$ is a **triple proportion**. It can also be expressed as $a: c: e = b: d: f$.

(4) In Notes (2) and (3), the real number k is called the **constant of proportionality**.

Example 7.4: The proportion $a: b: c$ = 3:5:7 has constant of proportionality 4. What is the value of c?

Solution: We are given that $\frac{a}{3} = \frac{b}{5} = \frac{c}{7} = 4$. So $c = 7 \cdot 4 = \mathbf{28}$.

Example 7.5: Two numbers are in the ratio 2:3 and their sum is 50. Evaluate the product of these two numbers.

Solution: Since the two numbers are in the ratio 2:3, we can represent them by $2x$ and $3x$. So $2x + 3x = 50$, or equivalently, $5x = 50$. So $x = 10$, and the two numbers are $2x = 2 \cdot 10 = 20$ and $3x = 3 \cdot 10 = 30$.

So the product of the two numbers is $20 \cdot 30 = \mathbf{600}$.

Example 7.6: Two positive integers are in the ratio 17:15. What is the least possible value for the sum of these two integers?

Solution: Since the two numbers are in the ratio 17:15, we can represent them by $17x$ and $15x$. Letting $x = 1$, we see that the least possible value for these integers is 17 and 15, so that their sum is $17 + 15 = \mathbf{32}$.

Example 7.7: If $\frac{a+b}{a} = 7$, then $\frac{a+b}{b} =$

Solution: Since $\frac{a+b}{a} = 7$, we have $a + b = 7a$, or $b = 7a - a = 6a$. So $\frac{a+b}{b} = \frac{a+6a}{6a} = \frac{7a}{6a} = \mathbf{\frac{7}{6}}$.

Example 7.8: If $\frac{6}{x} = \frac{12}{34}$, then $x =$

Solution: Since $\frac{6}{x} = \frac{12}{34}$, we have $12x = 6 \cdot 34$, and so $x = \frac{6 \cdot 34}{12} = \mathbf{17}$.

Example 7.9: When 8 kg of fresh apricots are dried, we obtain 5 kg of dried apricots. How many fresh apricots would we need to order in order to obtain 20 kg of dried apricots?

Solution: We identify 2 key words that tell us what 2 things are being compared. In this case, such a pair of key words is "fresh" and "dried."

fresh	8	x
dried	5	20

Now draw in the division symbols and equal sign, cross multiply and divide the corresponding ratio to find the unknown quantity x.

$$\frac{8}{5} = \frac{x}{20} \Longrightarrow 5x = 8 \cdot 20 \Longrightarrow x = \frac{8 \cdot 20}{5} = \mathbf{32}$$

DIRECT PROPORTIONALITY

The following are all equivalent ways of saying the same thing:

(1) y varies directly as x

(2) y is directly proportional to x

(3) $y = kx$ for some constant k

(4) $\frac{y}{x}$ is constant

(5) the graph of $y = f(x)$ is a nonvertical line through the origin.

For example, in the equation $y = 5x$, y varies directly as x. Here is a partial table of values for this equation.

x	1	2	3	4
y	5	10	15	20

Note that we can tell that this table represents a direct relationship between x and y because $\frac{5}{1} = \frac{10}{2} = \frac{15}{3} = \frac{20}{4}$. Here the **constant of proportionality** is 5.

Here is a graph of the equation.

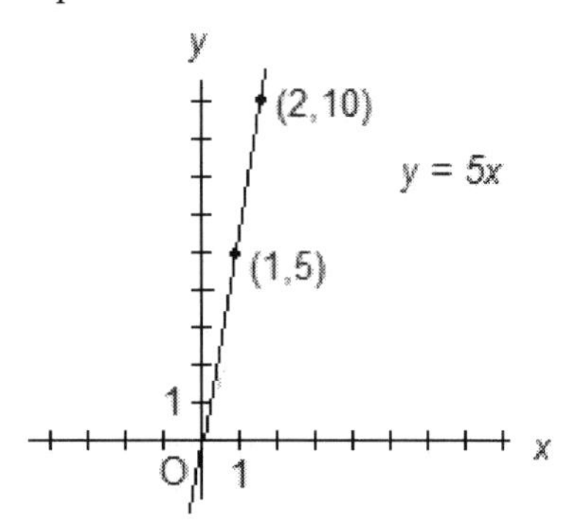

Note that we can tell that this graph represents a direct relationship between x and y because it is a nonvertical line through the origin. The constant of proportionality is the slope of the line, in this case $m = 5$.

Example 7.10: If y varies directly as x and $y = 5$ when $x = 8$, then what is y when $x = 24$?

Solution 1: Since y varies directly as x, $y = kx$ for some constant k. We are given that $y = 5$ when $x = 8$, so that $5 = k(8)$, or $k = \frac{5}{8}$. Therefore $y = \frac{5x}{8}$. When $x = 24$, we have $y = \frac{5(24)}{8} = \mathbf{15}$.

Solution 2: Since y varies directly as x, $\frac{y}{x}$ is a constant. So we get the following ratio: $\frac{5}{8} = \frac{y}{24}$. Cross multiplying gives $120 = 8y$, so that $y = \mathbf{15}$.

Solution 3: The graph of $y = f(x)$ is a line passing through the points $(0, 0)$ and $(8, 5)$. The slope of this line is $\frac{5-0}{8-0} = \frac{5}{8}$. Writing the equation of the line in slope-intercept form we have $y = \frac{5}{8}x$. As in solution 1, when $x = 24$, $y = \frac{5(24)}{8} = \mathbf{15}$.

Solution 4: To get from $x = 8$ to $x = 24$ we multiply x by 3. So we have to also multiply y by 3. We get $3(5) = \mathbf{15}$.

INVERSE PROPORTIONALITY

The following are all equivalent ways of saying the same thing:

(1) y varies inversely as x
(2) y is inversely proportional to x
(3) $y = \frac{k}{x}$ for some constant k
(4) xy is constant

The following is a consequence of (1), (2) (3) or (4).

(5) The graph of $y = f(x)$ is a hyperbola.

Note: (5) is not equivalent to (1), (2), (3) or (4).

For example, in the equation $y = \frac{12}{x}$, y varies inversely as x. Here is a partial table of values for this equation.

x	1	2	3	4
y	12	6	4	3

Note that we can tell that this table represents an inverse relationship between x and y because (1)(12) = (2)(6) = (3)(4) = (4)(3) = 12. Here the **constant of proportionality** is 12.

Here is a graph of the equation. On the left you can see the full graph. On the right we have a close-up in the first quadrant.

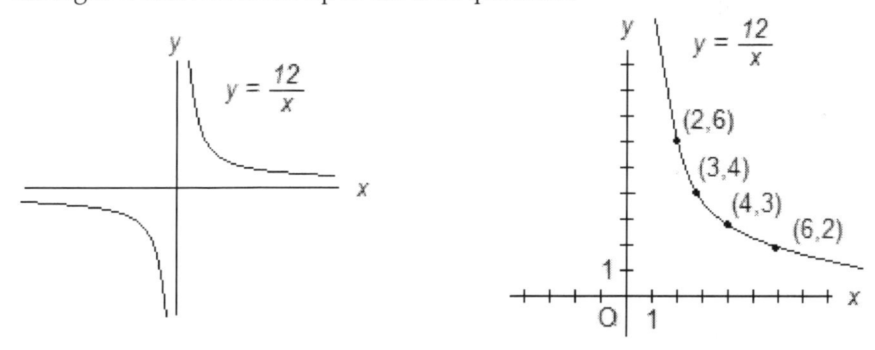

Example 7.11: If y varies inversely as x and $y = 8$ when $x = 3$, then what is y when $x = 6$?

Solution 1: Since y varies inversely as x, $y = \frac{k}{x}$ for some constant k. We are given $y = 8$ when $x = 3$, so that $8 = \frac{k}{3}$, or $k = 24$. Thus, $y = \frac{24}{x}$. When $x = 6$, we have $y = \frac{24}{6} = \mathbf{4}$.

Solution 2: Since y varies inversely as x, xy is a constant. So we get the following equation: $(3)(8) = 6y$ So $24 = 6y$, and $y = \frac{24}{6} = \mathbf{4}$.

Solution 3: $\frac{(8)(3)}{6} = \mathbf{4}$.

Example 7.12: If 6 workers can carpet a building in 16 days, then in how many days can 12 workers carpet the same building?

Solution: There is an inverse relationship between the number of workers and the number of days needed to complete the work. Modelling solution 3 in example 7.11, we get $\frac{(6)(16)}{12} = 8$. So 12 workers can carpet the building in **8 days**.

Example 7.13: The perimeter of a rectangle is 320 cm, and unequal sides have lengths in the ratio 3:5. Calculate the length of a shorter side of the rectangle.

Solution: Since the side lengths are in the ratio 3:5, we can represent them by $3x$ and $5x$. Since the perimeter is 320, we get $2(3x + 5x) = 320$, or equivalently $16x = 320$. So $x = \frac{320}{16} = 20$, and therefore the length of a shorter side of the rectangle is $3x = 3 \cdot 20 =$ **60 cm**.

8. EXPONENTS

Definition: If a is any real number and n is a positive integer, a^n means to multiply the number a by itself n times.

In the expression a^n, a is called the **base**, and n is called the **exponent**.

$$a^n = \underbrace{a.a.a\ldots\ldots a}_{n}$$

Example 8.1: $2^1 = 2$

$2^2 = 2 \cdot 2 = 4$

$2^3 = 2 \cdot 2 \cdot 2 = 8$

$3^4 = 3 \cdot 3 \cdot 3 \cdot 3 = 81$

$x^5 = x \cdot x \cdot x \cdot x \cdot x$

Note: Raising any number to the power of 1 always gives the number we started with. That is, for any real number x, we always have $x^1 = x$.

Also, by definition, raising any nonzero real number to the power of zero always gives 1. In other words, for any nonzero real number x, we have $x^0 = 1$.

Example 8.2: Compute $2^3 \cdot 2^4$

Solution: $2^3 = 2 \cdot 2 \cdot 2$ and $2^4 = 2 \cdot 2 \cdot 2 \cdot 2$. So $2^3 \cdot 2^4$

$$= (2 \cdot 2 \cdot 2)(2 \cdot 2 \cdot 2 \cdot 2) = 2^7 = \mathbf{128}.$$

Note that we just discovered that $2^3 \cdot 2^4 = 2^7$.

This discovery leads us to the following rule:

When multiplying two expressions with the same base, we can add the exponents.

In other words, we have $\boldsymbol{x^a x^b = x^{a+b}}$.

Example 8.3: Compute $\frac{7^8}{7^6}$

Solution: $7^8 = 7 \cdot 7 \cdot 7 \cdot 7 \cdot 7 \cdot 7 \cdot 7 \cdot 7$ and $7^6 = 7 \cdot 7 \cdot 7 \cdot 7 \cdot 7 \cdot 7$.

So $\frac{7^8}{7^6} = \frac{7 \cdot 7 \cdot 7 \cdot 7 \cdot 7 \cdot 7 \cdot 7 \cdot 7}{7 \cdot 7 \cdot 7 \cdot 7 \cdot 7 \cdot 7} = 7 \cdot 7 = \mathbf{49}$.

Note that we just discovered that $\frac{7^8}{7^6} = 7^2$.

This discovery leads us to the following rule:

When dividing two expressions with the same base, we can subtract the exponents.

In other words, we have $\boldsymbol{\frac{x^a}{x^b} = x^{a-b}}$.

Example 8.4: Express $(x^3)^5$ using only one exponent.

Solution: $(x^3)^5 = x^3 \cdot x^3 \cdot x^3 \cdot x^3 \cdot x^3$

$$= (x \cdot x \cdot x)(x \cdot x \cdot x)(x \cdot x \cdot x)(x \cdot x \cdot x)(x \cdot x \cdot x) = \boldsymbol{x^{15}}.$$

Note that we just discovered that $(x^3)^5 = x^{15}$.

When raising an expression to an exponent, then to a second exponent, we can multiply the exponents.

In other words, we have $\boldsymbol{(x^a)^b = x^{ab}}$.

We summarize these rules (and two more) in the following table:

Law	Example
$x^0 = 1$	$3^0 = 1$
$x^1 = x$	$9^1 = 9$
$x^a x^b = x^{a+b}$	$x^3 x^5 = x^8$
$x^a / x^b = x^{a-b}$	$x^{11} / x^4 = x^7$
$(x^a)^b = x^{ab}$	$(x^5)^3 = x^{15}$
$(xy)^a = x^a y^a$	$(xy)^4 = x^4 y^4$
$(x/y)^a = x^a / y^a$	$(x/y)^6 = x^6 / y^6$

Example 8.5: $(424)^0 = 1$ $\quad$ $1^0 = 1$ $\quad$ $1^{324} = 1$

NEGATIVE AND FRACTIONAL EXPONENTS

By definition, we have $\boldsymbol{x^{-1} = \frac{1}{x}}$

Example 8.6: $5^{-1} = \frac{1}{5}$ or **.2**.

If x is a nonzero real number, and n is a positive integer, we have

$$\boldsymbol{x^{-n} = (x^n)^{-1} = \frac{1}{x^n}}$$

Note: Here we have used a law of exponents together with the definition of x^{-1}.

Example 8.7: $3^{-4} = \frac{1}{3^4} = \boldsymbol{\frac{1}{81}}$.

By definition, we have $\boldsymbol{x^{\frac{1}{n}} = \sqrt[n]{x}}$

Example 8.8: $8^{1/4} = \boldsymbol{\sqrt[4]{8}}$.

We summarize these rules (and one more) in the following table:

Law	Example
$\boldsymbol{x^{-1} = \frac{1}{x}}$	$3^{-1} = \frac{1}{3}$
$\boldsymbol{x^{-a} = \frac{1}{x^a}}$	$9^{-2} = \frac{1}{81}$
$\boldsymbol{x^{\frac{1}{n}} = \sqrt[n]{x}}$	$x^{\frac{1}{3}} = \sqrt[3]{x}$
$\boldsymbol{x^{\frac{m}{n}} = \sqrt[n]{x^m} = \left(\sqrt[n]{x}\right)^m}$	$x^{\frac{9}{2}} = \sqrt{x^9} = \left(\sqrt{x}\right)^9$

EXPONENTIAL EQUATIONS

Rules: If $a^m = a^n$ then $m = n$.

If $a^m = b^m$ then $\begin{cases} a = b & \text{if } m \text{ is odd} \\ a = b \text{ or } a = -b & \text{if } m \text{ is even} \end{cases}$

Example 8.9: If $3^x = 27$, what is the value of x ?

Solution: $3^x = 27 \Rightarrow 3^x = 3^3 \Rightarrow x = \mathbf{3}$.

Example 8.10: $x^3 = 64 \Rightarrow x^3 = 4^3 \Rightarrow x = \mathbf{4}$.

Example 8.11: $x^4 = 81 \Rightarrow x^4 = 3^4 \Rightarrow x = \mathbf{3}$.

Example 8.12: $\frac{3^x}{7^x} = \frac{9}{49} \Rightarrow \frac{3^x}{7^x} = \frac{3^2}{7^2} \Rightarrow \left(\frac{3}{7}\right)^x = \left(\frac{3}{7}\right)^2 \Rightarrow x = \mathbf{2}$.

Example 8.13: Compute $\left(-\frac{1}{2}\right)^2 + \left(\frac{1}{2}\right)^2$.

Solution: $\left(-\frac{1}{2}\right)^2 + \left(\frac{1}{2}\right)^2 = \frac{1}{4} + \frac{1}{4} = \frac{2}{4} = \boldsymbol{\frac{1}{2}}$.

Example 8.14: Compute $3^{-2} + 10^{-2}$.

Solution: $3^{-2} + 10^{-2} = \frac{1}{3^2} + \frac{1}{10^2} = \frac{1}{9} + \frac{1}{100} = \frac{100}{900} + \frac{9}{900} = \mathbf{\frac{109}{900}}$.

Example 8.15: Compute $(-3)^{-2} + (-5)^{-2}$.

Solution: $(-3)^{-2} + (-5)^{-2} = \frac{1}{(-3)^2} + \frac{1}{(-5)^2} = \frac{1}{9} + \frac{1}{25} = \frac{25+9}{9 \cdot 25} = \mathbf{\frac{34}{225}}$.

POSITIVE AND NEGATIVE POWERS OF 10

$10^1 = 10$ $\quad 10^{-1} = \frac{1}{10} = .1$

$10^2 = 100$ $\quad 10^{-2} = \frac{1}{100} = .01$

$10^3 = 1000$ $\quad 10^{-3} = \frac{1}{1000} = .001$

$10^4 = 10{,}000$ $\quad 10^{-4} = \frac{1}{10{,}000} = .0001$

Example 8.16: Express 26.426 in expanded form.

Solution: $26.426 = \mathbf{2 \times 10^1 + 6 \times 10^0 + 4 \times 10^{-1} + 2 \times 10^{-2} + 6 \times 10^{-3}}$.

Example 8.17: Express 208.506 in expanded form.

Solution:

$\mathbf{208.506 = 2 \times 10^2 + 0 \times 10^1 + 8 \times 10^0 + 5 \times 10^{-1} + 0 \times 10^{-2} + 6 \times 10^{-3}}$.

Example 8.18: Write $\frac{3^{93} - 3^{92}}{3^{94}}$ as a simple fraction without exponents.

Solution: $\frac{3^{93} - 3^{92}}{3^{94}} = \frac{3^{92}(3-1)}{3^{92+2}} = \frac{3^{92} \cdot 2}{3^{92} \cdot 3^2} = \mathbf{\frac{2}{9}}$.

Example 8.19: Write $\frac{7^2 + (-7)^2}{(-1)^6 + 7^2}$ as a simple fraction without exponents.

Solution: $\frac{7^2 + (-7)^2}{(-1)^6 + 7^2} = \frac{49+49}{1+49} = \frac{98}{50} = \mathbf{\frac{49}{25}}$.

Example 8.20: Compute $\frac{1}{10}$ of 10^{10}.

Solution: $\frac{1}{10} \cdot 10^{10} = \frac{10^{10}}{10^1} = \mathbf{10^9}$.

Example 8.21: If $2^x = a$ and $3^x = b$, then in terms of a and b, $24^x =$

Solution: $24^x = (8 \cdot 3)^x = 8^x \cdot 3^x = (2^3)^x \cdot 3^x = (2^x)^3 \cdot 3^x = \boldsymbol{a^3 b}$.

Example 8.22: Evaluate $\frac{7^n + 7^{n+1}}{7^n + 7^n}$.

Solution: $\frac{7^n + 7^{n+1}}{7^n + 7^n} = \frac{7^n + 7^n \cdot 7^1}{2 \cdot 7^n} = \frac{7^n(1+7)}{7^n \cdot 2} = \frac{8}{2} = \mathbf{4}$.

Example 8.23: Write $\frac{11^{12}+11^{13}}{11^{13}-11^{14}}$ as a simple fraction without exponents.

Solution: $\frac{11^{12}+11^{13}}{11^{13}-11^{14}} = \frac{11^{12}+11^{12+1}}{11^{13}-11^{13+1}} = \frac{11^{12}+11^{12}\cdot 11^{1}}{11^{13}-11^{13}\cdot 11^{1}} = \frac{11^{12}(1+11)}{11^{13}(1-11)} = \frac{12}{11(-10)} = -\frac{\mathbf{6}}{\mathbf{55}}$.

Example 8.24: If $2 \cdot 17^{14} + 6 \cdot 17^{14} + 9 \cdot 17^{14} = 17^{n}$, then $n =$

Solution:

$$2 \cdot 17^{14} + 6 \cdot 17^{14} + 9 \cdot 17^{14} = (2 + 6 + 9)17^{14} = 17 \cdot 17^{14} = 17^{15}.$$

So $n = \mathbf{15}$.

Example 8.25: If $(7^{-4})^{3} \cdot (7^{-3})^{4} = 7^{n}$, then $n =$

Solution: $(7^{-4})^{3} \cdot (7^{-3})^{4} = 7^{-12} \cdot 7^{-12} = 7^{-24}$. So $n = \mathbf{-24}$.

Example 8.26: $(-3)^{0} + (-3)^{1} + (-3)^{2} + (-3)^{3} =$

Solution: $(-3)^{0} + (-3)^{1} + (-3)^{2} + (-3)^{3} = 1 - 3 + 9 - 27 = \mathbf{-20}$.

9. PERCENTAGES

The word "percent" means "out of 100." We use the symbol "%" for percent. So, for example, 5% means "5 out of 100," or equivalently $\frac{5}{100} = \frac{1}{20}$.

- In general, to change a percent to a fraction, simply place the percentage over 100. You may then want to reduce the fraction, as was done with 5% above.
- To change a percent to a decimal, divide by 100, or equivalently move the decimal point two places to the left (adding zeros if necessary). So, for example, 5% = .05.
- To change a fraction to a percent, set up a ratio where the second fraction has denominator 100. Then cross multiply and divide. For example, we can change the fraction $\frac{9}{30}$ to a percent as follows:

$$\frac{9}{30} = \frac{x}{100} \Rightarrow 30x = 900 \Rightarrow x = \frac{900}{30} = 30 \Rightarrow \frac{9}{30} = 30\%.$$

- To change a decimal to a percent, multiply by 100, or equivalently move the decimal point two places to the right (adding zeros if necessary). So, for example, .23 = 23%.

Here are some more examples:

$15\% = .15 = \frac{15}{100} = \frac{3}{20}$ $20\% = .2 = \frac{20}{100} = \frac{1}{5}$

$45\% = .45 = \frac{45}{100} = \frac{9}{20}$ $150\% = 1.5 = \frac{150}{100} = \frac{3}{2}$

Solving problems involving percentages usually involve evaluating an unknown by using two given values among "base" (B), "rate" (R), and "portion" (P).

Assume that, $x\%$ of Δ is P.

Here, Δ = "base"(B), x = "rate"(R), and P = "portion"(P).

$$\frac{x}{100} \cdot \Delta = P$$

As an alternative, by moving Δ to the denominator of the right hand side, we can think of this as the following ratio:

$$\frac{x}{100} = \frac{P}{\Delta}$$

Note that the base Δ represents the total and P is the portion of the total that we would like the percentage of.

"x% of Δ" $= \frac{x}{100} \cdot \Delta$

Example 9.1: What is 20% of 30?

Solution: "20% of 30" $= \frac{20}{100} \cdot 30 = \mathbf{6}$.

Example 9.2: Evaluate 8% of 8?

Solution: "8% of 8" $= \frac{8}{100} \cdot 8 = \mathbf{\frac{16}{25}}$.

Example 9.3: What is 75% of 600?

Solution: "75% of 600" $= \frac{75}{100} \cdot 600 = \mathbf{450}$.

Example 9.4: What percentage of 400 equals to 20?

Solution: $\frac{x}{100} \cdot 400 = 20 \Rightarrow x = 20 \cdot \frac{100}{400} = 5$.

So the answer is **5**%.

"x% more than Δ" $= \frac{100+x}{100} \cdot \Delta$

"x% less than Δ" $= \frac{100-x}{100} \cdot \Delta$

Example 9.5: Evaluate 40% more than 200.

Solution: $\frac{100+40}{100} \cdot 200 = \frac{140}{100} \cdot 200 = \mathbf{280}.$

Alternative solution: $200 + .40 \cdot 200 = 200 + 80 = \mathbf{280}.$

Example 9.6: Calculate 60% less than 600.

Solution: $\frac{100-60}{100} \cdot 600 = 40 \cdot 6 = \mathbf{240}.$

Alternative solution: $600 - .60 \cdot 600 = 600 - 360 = \mathbf{240}.$

Example 9.7: Calculate 20% more than 10% more than x.

Solution: $\frac{100+20}{100} \cdot \frac{100+10}{100} \cdot x = \frac{120}{100} \cdot \frac{110}{100} \cdot x = \mathbf{1.32}\boldsymbol{x}.$

Example 9.8: Calculate 20% less than 10% more than x.

Solution: $\frac{100-20}{100} \cdot \frac{100+10}{100} \cdot x = \frac{80}{100} \cdot \frac{110}{100} \cdot x = \mathbf{.88}\boldsymbol{x}.$

Example 9.9: When the number A is increased by 40%, the result is 280. What is the value of A ?

Solution: $\frac{140}{100}A = 280 \Rightarrow A = 280 \cdot \frac{100}{140} \Rightarrow A = \mathbf{200}.$

Example 9.10: When a certain number is increased by 70%, the result is 340. Find the number.

Solution: Let x be the number. Then $\frac{170}{100}x = 340 \Rightarrow x = 340 \cdot \frac{100}{170} = \mathbf{200}.$

Example 9.11: Calculate the sum of 20% of 60 and 30% of 60.

Solution: $.2 \cdot 60 + .3 \cdot 60 = .5 \cdot 60 = \mathbf{30}.$

Example 9.12: Calculate the positive difference between 30% of 80 and 20% of 80.

Solution: $.3 \cdot 80 - .2 \cdot 80 = .1 \cdot 80 = \mathbf{8}.$

Example 9.13: If 20% of x is .06, then what is the value of x?

Solution: $.2x = .06 \Rightarrow x = \frac{.06}{.2} = \frac{6}{20} = \mathbf{\frac{3}{10}}$ or $\mathbf{.3}$.

Example 9.14: If 25% of a number is $\sqrt{75}$, then what is the value of this number?

Solution: Let x be the number. Then

$$.25x = \sqrt{75} \Rightarrow x = \frac{\sqrt{75}}{.25} = \frac{\sqrt{25}\sqrt{3}}{.25} = \frac{5\sqrt{3}}{\frac{1}{4}} = 4 \cdot 5\sqrt{3} = \mathbf{20\sqrt{3}}.$$

Example 9.15: A book was purchased by a retailer for \$60 and sold at a 30% profit. What was the sale price of the book?

Solution: The profit was $.3 \cdot 60 = \$18$. It follows that the sale price was $60 + 18 = \mathbf{\$78}$.

Example 9.16: A mobile phone was purchased by a retailer for $300 and sold at a 40% loss. What was the sale price of the phone?

Solution: The loss was $.4 \cdot 300 = \$120$. It follows that the sale price was $300 - 120 = \mathbf{\$180}$.

SIMPLE INTEREST

$I = Prt$, where I is the interest earned on P dollars at an annual rate of r% (written as a fraction or decimal) for t years.

Note: P stands for **principal** which is the amount of money invested.

r stands for **rate**. This is the annual interest rate written as a fraction or decimal.

t stands for **time** in years.

I stands for **interest**.

Example 9.17: $600 was invested, and after 8 months the amount of interest earned was $30. What was the annual interest rate?

Solution: We are given that $P = 600$, $t = \frac{8}{12} = \frac{2}{3}$, and $I = 30$. So we have $30 = 600 \cdot r \cdot \frac{2}{3} = 400r$. So the annual interest rate was

$$r = \frac{30}{400} = .075 = \mathbf{7.5\%}.$$

Example 9.18: $2000 is deposited into an account for 6 years with an annual interest rate of 6%. After 6 years, all of the money will be withdrawn from bank. What is the total amount of money that will be withdrawn?

Solution: We are given that $P = 2000$, $t = 6$, and $r = .06$. So we have $I = 2000 \cdot .06 \cdot 6 = 720$. The total amount of money withdrawn is then $A = P + I = 2000 + 720 = \mathbf{2720}$.

Example 9.19: The cost of a certain product is $160. If we want to sell the product with 20% interest, what should the selling price be?

Solution:

$$\text{Selling price} = \text{Cost} + \text{Interest} = 160 + 160 \cdot .2 = 1.2 \cdot 160 = \mathbf{\$192}.$$

Example 9.20: The selling price of an item is discounted 30% over its initial ticket price of $500. A second discount of 20% is then applied to the already discounted price. What will be the final selling price of the item?

Solution: After the first discount is applied, the new selling price would be $500 - 500 \cdot .3 = .7 \cdot 500 = 350$.

After the second discount is applied, the final selling price would be $350 - 350 \cdot .2 = .8 \cdot 350 = \$\mathbf{280}$.

Example 9.21: If $600 is deposited into a bank account bearing 3% interest, what will be the total amount of money in the account after 1 year?

Solution: Total = Principal + Annual interest = $600 + 600 \cdot .03 = 1.03 \cdot 600 = \$\mathbf{618}$.

Example 9.22: $400 is deposited into a bank account with an annual interest rate of 3%. What is the amount of interest earned after 4 months?

Solution: $I = Prt = 400 \cdot .03 \cdot \frac{4}{12} = \$\mathbf{4}$.

Example 9.23: A salesman sells 60% of his goods, and then sells 25% of his remaining goods. What percentage of his initial goods did he sell?

Solution: If we assume the salesman started with 100 goods, then after he sells 60% of his goods, he is left with 100 – 60 = 40 goods. After selling 25% of these 40 goods, he is left with $40 - .25 \cdot 40 = 40 - 10 = 30$ goods. So he now has 30 goods, which means that he sold 70 of his hundred goods. So he sold **70%** of his initial goods.

Note: One technique for solving percent problems is to start with the number 100. After all, the word "percent" means "out of 100."

Formal solution: If the salesman starts with x goods, then after selling 60% of these goods, he is left with $x - .6x = (1 - .6)x = .4x$ goods. After selling 25% of these $.4x$ goods, he will now be left with

$$.4x - .25 \cdot .4x = .4x - .1x = .3x \text{ goods.}$$

So he now has $.3x$ goods, which means that he sold $x - .3x = .7x$ goods. So he sold **70%** of his initial goods.

10. FACTORING

GCD Factoring: (i) We find the GCD of all the terms in the expression, (ii) we divide each term by this GCD, and then (iii) multiply the resulting expression by the GCD.

Example 10.1: $3a + 9b = 3(a + 3b)$

Breakdown: (i) GCD($3a$,$9b$) = 3.

(ii) $\frac{3a}{3} = a, \frac{9b}{3} = 3b$. So after the division we have $a + 3b$.

(iii) We multiply $a + 3b$ by 3 to get $3(a + 3b)$.

Example 10.2: $4a + 8ab = 4a(1 + 2b)$

Example 10.3: $mn - mk = m(n - k)$

Example 10.4: $9a - 3a^2 = 3a(3 - a)$

Example 10.5: $mn^2 - nm^2 = mn(n - m)$

Example 10.6: $6a + 6b + 9c = 3(2a + 2b + 3c)$

Example 10.7: $8x^2 + 6x = 2x(4x + 3)$

Example 10.8: $4m + 16m^2 = 4m(1 + 4m)$

Example 10.9: $3mn + 9m^2n^2 = 3mn(1 + 3mn)$

Factoring by Grouping: (i) Terms are placed into groups of two, three or more, (ii) we factor each group separately, and then (iii) we factor across groups.

Example 10.10: $mn + 3n + 3m + 9 = (mn + 3n) + (3m + 9)$

$$= n(m + 3) + 3(m + 3) = (m + 3)(n + 3)$$

Breakdown: (i) We group the first two terms together and the last two terms together: $(mn + 3n) + (3m + 9)$

(ii) We factor each group: $mn + 3n = n(m + 3)$ and $3m + 9 = 3(m + 3)$

(iii) We factor across the two groups, noting that the GCD of the two groups is $m + 3$. So we get $n(m + 3) + 3(m + 3) = (m + 3)(n + 3)$.

Example 10.11: $m^2n - 4mn + 8m - 32 = (m^2n - 4mn) + (8m - 32)$

$$= mn(m - 4) + 8(m - 4) = (m - 4)(mn + 8)$$

Example 10.12: $5a^3 - 10a^2 + 4a - 8 = (5a^3 - 10a^2) + (4a - 8)$

$$= 5a^2(a - 2) + 4(a - 2) = (a - 2)(5a^2 + 4)$$

Factoring the Difference of Two Squares: $(a^2 - b^2) = (a + b)(a - b)$

Example 10.13: $9 - m^2 = (3 + m)(3 - m)$

Example 10.14: $m^2 - 16 = (m + 4)(m - 4)$

Example 10.15: $25m^2 - 4n^2 = (5m + 2n)(5m - 2n)$

Example 10.16: $1 - m^2 = (1 + m)(1 - m)$

Example 10.17: $144x^2 - 169n^2 = (12x + 13n)(12x - 13n)$

Example 10.18: $m^4 - n^4 = (m^2 + n^2)(m^2 - n^2)$

$$= (m^2 + n^2)(m + n)(m - n)$$

Example 10.19: $m - n = (\sqrt{m} + \sqrt{n})(\sqrt{m} - \sqrt{n})$

Example 10.20: $\frac{a^2}{16} - \frac{b^2}{4} = \left(\frac{a}{4}\right)^2 - \left(\frac{b}{2}\right)^2 = \left(\frac{a}{4} + \frac{b}{2}\right)\left(\frac{a}{4} - \frac{b}{2}\right)$

Factoring Perfect Squares:

$$a^2 + 2ab + b^2 = (a + b)^2 \qquad a^2 - 2ab + b^2 = (a - b)^2$$

Note: The sign in front of b^2 must be "+," but the sign in front of $2ab$ can be "+" or "−."

Example 10.21: $4a^2 + 12ab + 9b^2 = (2a + 3b)^2$

Breakdown: (i) The square root of $4a^2$ is $2a$ and the square root of $9b^2$ is $3b$.

(ii) Note that $2 \cdot 2a \cdot 3b = 12ab$. So the result is $(2a + 3b)^2$

Example 10.22: $64a^2 - 48ab + 9b^2 = (8a - 3b)^2$

Example 10.23: If $49a^2 + nab + 16b^2$ is a perfect square, then what is the value of n ?

Solution: We must have $nab = 2(7a)(4b) = 56ab$. So $n = \mathbf{56}$.

Example 10.24: $(x + 4)^2 = x^2 + 2 \cdot 4 \cdot x + 4^2 = x^2 + 8x + 16$

Example 10.25: $(x + 5)^2 = x^2 + 2 \cdot 5 \cdot x + 5^2 = x^2 + 10x + 25$

Example 10.26: $(2x + 3)^2 = (2x)^2 + 2 \cdot 3 \cdot 2x + 3^2 = 4x^2 + 12x + 9$

Example 10.27: $(3x + 3)^2 = (3x)^2 + 2 \cdot 3 \cdot 3x + 3^2 = 9x^2 + 18x + 9$

Example 10.28: Simplify $\left(\sqrt{6} + \sqrt{3}\right)^2 - (\sqrt{6} - \sqrt{3})$?

Solution: $\left(\sqrt{6} + \sqrt{3}\right)^2 - \left(\sqrt{6} - \sqrt{3}\right) = \left(6 + 2\sqrt{6}\sqrt{3} + 3\right) - (\sqrt{6} - \sqrt{3})$

$$= 9 + 2\sqrt{18} - \sqrt{6} - \sqrt{3} = \mathbf{9 + 6\sqrt{2} - \sqrt{6} - \sqrt{3}}.$$

Example 10.29: $(x - 2)^2 = x^2 - 2 \cdot 2 \cdot x + 2^2 = x^2 - 4x + 4$

Example 10.30: $(x - 7)^2 = x^2 - 2 \cdot 7 \cdot x + 7^2 = x^2 - 14x + 49$

Example 10.31: $(2x - 5)^2 = (2x)^2 - 2 \cdot 5 \cdot 2x + 5^2 = 4x^2 - 20x + 25$

Example 10.32: $(4y - 3)^2 = (4y)^2 - 2 \cdot 3 \cdot 4y + 3^2 = 16y^2 - 24y + 9$

Factoring Trinomials of the form $x^2 + bx + c$:

$$x^2 + (m + n)x + mn = (x + m)(x + n)$$

Example 10.33: $x^2 + 7x + 6 = (x + 1)(x + 6)$

Breakdown: (i) The product of the first terms of each factor must be x^2. So each of these terms must be x: $x^2 + 7x + 6 = (x + \_)(x + \_)$.

(ii) The product of the last terms of each factor must be 6. There are two choices: 1 and 6, or 2 and 3.

(iii) The sum of the last terms of each factor must be 7. Since 1 + 6 = 7, the two factors must be 1 and 6. So we get $(x + 1)(x + 6)$.

Example 10.34: $x^2 + 6x + 8 = (x + 4)(x + 2)$

$$x^2 - (m + n)x + mn = (x - m)(x - n)$$

Example 10.35: $x^2 - 10x + 21 = (x - 7)(x - 3)$

$$x^2 + (m - n)x - mn = (x + m)(x - n)$$

Example 10.36: $x^2 + 3x - 28 = (x + 7)(x - 4)$

Example 10.37: If $x = \sqrt{6} - 8$, what is the value of $x^2 + 16x + 64$?

Solution: $x^2 + 16x + 64 = (x + 8)^2 = (\sqrt{6} - 8 + 8)^2 = \left(\sqrt{6}\right)^2 = \mathbf{6}$.

Factoring Trinomials of the form $ax^2 + bx + c$:

Example 10.38: $3x^2 + 10x + 8 = (3x + 4)(x + 2)$

Breakdown: (i) The product of the first terms of each factor must be $3x^2$. So these terms must be $3x$ and x: $3x^2 + 10x + 8 = (3x + \_)(x + \_)$.

(ii) The product of the last terms of each factor must be 8. There are four choices: 1 and 8, 2 and 4, 4 and 2, or 8 and 1.

Note that this time there is no symmetry. For example 2 and 4 is genuinely different from 4 and 2.

$$(3x + 2)(x + 4) = 3x^2 + 12x + 2x + 8 = 3x^2 + 14x + 8$$

$$(3x + 4)(x + 2) = 3x^2 + 6x + 4x + 8 = 3x^2 + 10x + 8$$

(iii) Note that 4 and 2 is the correct choice because

$$3x \cdot 2 + 4 \cdot x = 6x + 4x = 10x.$$

The following figure might be useful.

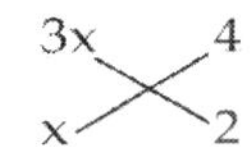

Example 10.39: Factor $3x^2 + 19x + 20$

Solution:

3x 4

x 5

$3x^2 + 19x + 20 = (3x + 4)(x + 5)$

Example 10.40: Evaluate $3001 \cdot 3004 - 3002 \cdot 3003$.

Solution: Let $a = 3001$. Then we have $3002 = a + 1$, $3003 = a + 2$, and $3004 = a + 3$. So $3001 \cdot 3004 - 3002 \cdot 3003 = a(a + 3) - (a + 1)(a + 2)$

$$= a^2 + 3a - (a^2 + 2a + a + 2) = a^2 + 3a - a^2 - 2a - a - 2 = \mathbf{-2}.$$

Example 10.41: If $x + \frac{1}{x} = 7$, then $x^2 + \frac{1}{x^2} =$

Solution: $\left(x + \frac{1}{x}\right)^2 = x^2 + 2x \cdot \frac{1}{x} + \left(\frac{1}{x}\right)^2 = x^2 + 2 + \frac{1}{x^2}$.

So $x^2 + \frac{1}{x^2} = \left(x + \frac{1}{x}\right)^2 - 2 = 7^2 - 2 = 49 - 2 = \mathbf{47}$.

Example 10.42: If a and b are positive integers with $a^2 - b^2 = 17$, then evaluate $a^2 + b^2$.

Solution: $(a + b)(a - b) = a^2 - b^2 = 17 = 17 \cdot 1$. So $a + b = 17$ and $a - b = 1$. Adding these equations gives the following.

$$\begin{array}{c} a + b = 17 \\ \underline{a - b = 1} \\ 2a = 18 \end{array}$$

So $a = \frac{18}{2} = 9$, and $b = 17 - a = 17 - 9 = 8$.

So $a^2 + b^2 = 9^2 + 8^2 = 81 + 64 = \mathbf{145}$.

Example 10.43: If $a = \sqrt{10} + 2$ and $b = \sqrt{10} - 2$, then $\frac{a}{b} + \frac{b}{a} =$

Solution:

$$\frac{a}{b} + \frac{b}{a} = \frac{a^2+b^2}{ab} = \frac{(\sqrt{10}+2)^2+(\sqrt{10}-2)^2}{(\sqrt{10}+2)(\sqrt{10}-2)} = \frac{(10+4\sqrt{10}+4)+(10-4\sqrt{10}+4)}{10-4} = \frac{28}{6} = \mathbf{\frac{14}{3}}.$$

Example 10.44: Factor $2ab + 2ac + 3db + 3dc$

Solution:

$$2ab + 2ac + 3db + 3dc = 2a(b + c) + 3d(b + c) = (\boldsymbol{b + c})(\boldsymbol{2a + 3d}).$$

Example 10.45: $m - 1 = \left(\sqrt{m}\right)^2 - (1)^2 = (\boldsymbol{\sqrt{m} + 1})(\boldsymbol{\sqrt{m} - 1})$

Example 10.46: $n - 5 = \left(\sqrt{n}\right)^2 - \left(\sqrt{5}\right)^2 = (\boldsymbol{\sqrt{n} + \sqrt{5}})(\boldsymbol{\sqrt{n} - \sqrt{5}})$

Example 10.47: $113^2 - 56^2 + 55^2 - 112^2 =$

Solution: $113^2 - 56^2 + 55^2 - 112^2$

$$= (113 + 56)(113 - 56) + (55 + 112)(55 - 112)$$

$$= 169 \cdot 57 + 167 \cdot (-57) = 57(169 - 167) = 57 \cdot 2 = \mathbf{114}.$$

Example 10.48: If $x^2 + xy = 55$ and $y^2 + xy = 66$, then find the positive value of $x + y$.

Solution:

$$\begin{array}{r} x^2 + xy = 55 \\ \underline{y^2 + xy = 66} \\ x^2 + y^2 + 2xy = 121 \end{array}$$

So $(x + y)^2 = x^2 + 2xy + y^2 = 121$, and $x + y = \pm 11$.

So the positive value of $x + y$ is **11**.

Example 10.49: If $x = 6.2$ and $y = 2.2$, evaluate $x^2 - 2xy + y^2$.

Solution: $x^2 - 2xy + y^2 = (x - y)^2 = (6.2 - 2.2)^2 = 4^2 = \mathbf{16}$.

Example 10.50: If $x = 7.3$ and $y = 4.7$, evaluate $(x - y)^2 + 4xy$.

Solution: $(x - y)^2 + 4xy = x^2 - 2xy + y^2 + 4xy$

$$= x^2 + 2xy + y^2 = (x + y)^2 = (7.3 + 4.7)^2 = 12^2 = \mathbf{144}.$$

Example 10.51: If $A^2 = \frac{1}{49} + \frac{1}{16} + \frac{1}{14}$, write each possible value of A as a simple fraction.

Solution: $\frac{1}{49} + \frac{1}{16} + \frac{1}{14} = \left(\frac{1}{7}\right)^2 + \left(\frac{1}{4}\right)^2 + 2 \cdot \frac{1}{7} \cdot \frac{1}{4} = \left(\frac{1}{7} + \frac{1}{4}\right)^2 = \left(\frac{4+7}{4 \cdot 7}\right)^2 = \left(\frac{11}{28}\right)^2$.

So $A^2 = \left(\frac{11}{28}\right)^2$ and $A = \pm\mathbf{\frac{11}{28}}$.

11. LINEAR EQUATIONS

LINEAR EQUATIONS WITH ONE UNKNOWN

Suppose we are asked to solve for x in the following equation:

$$x + 3 = 8$$

In other words, we are being asked for a number such that when we add 3 to that number we get 8. It is not too hard to see that $5 + 3 = 8$, so $x = 5$.

I call the technique above solving this equation **informally**. In other words, when we solve algebraic equations informally we are solving for the variable very quickly in our heads. I sometimes call this performing **"mental math."**

We can also solve for x **formally** by subtracting 3 from each side of the equation:

$$\begin{aligned} x + 3 &= 8 \\ -3 \ &\ -3 \\ \hline x \quad &= 5 \end{aligned}$$

In other words, when we solve an algebraic equation formally we are writing out all the steps – just as we would do it on a test in school. You should practice solving equations both informally and formally.

Example 11.1: Solve for x: $5x = 30$

Informal solution: 5 times 6 is 30, so we see that $x = \mathbf{6}$.

Formal solution: We can divide each side of the equation by 5:

$$\frac{5x}{5} = \frac{30}{5}$$

$$x = \mathbf{6}$$

Example 11.2: Solve for x: $5x + 3 = 48$

Informal solution: First let's figure out what number plus 3 is 48. Well, 45 plus 3 is 48. So $5x$ is 45. So x must be **9**.

Formal solution:

$$\begin{aligned} 5x + 3 &= 48 \\ -3 \ &\ -3 \\ \hline \frac{5x}{5} \quad &= \frac{45}{5} \\ x \quad &= \mathbf{9} \end{aligned}$$

Here are some basic rules for solving linear equations formally:

1. The same number can be added to or subtracted from each side of an equation.
2. Each side of an equation can be multiplied by the same number or divided by the same nonzero number.
3. A term can be moved to the other side of an equation by changing the sign in front of the term.
4. When solving a linear equation it is useful to move all constants to one side of the equation, and terms with variables to the other side of the equation.
5. In the linear equation $ax + b = 0$, a and b are called *coefficients* and x is called a *variable* or *unknown*.

Example 11.3: Find the value of x in the equation $6x-4=20$.

Solution: $6x-4=20 \Leftrightarrow 6x=24 \Leftrightarrow x=\mathbf{4}$.

Example 11.4: Find the value of x in the equation $4x-4=2x+2$.

Solution: $4x-4=2x+2 \Leftrightarrow 4x-2x=2+4 \Leftrightarrow 2x=6 \Leftrightarrow x=\mathbf{3}$.

Example 11.5: Solve for x: $3(x-10)=2(x+10)$

Solution: $3(x-10)=2(x+10) \Leftrightarrow 3x-30=2x+20 \Leftrightarrow x=\mathbf{50}$.

Notes: (1) Make sure you are using the **distributive property** correctly here. For example, $3(x-10)=3x+3(-10)=3x-30$. A common error is to write $3(x-10)=3x-10$.

(2) In general the distributive property says that for real number a, b, and c, we have

$$a(b+c)=ab+ac$$

(3) Observe that applying the distributive property is the same as GCF factoring. We are just performing the procedure in reverse.

Example 11.6: For what value of x does $\frac{2x-12}{5}=6$?

Solution: We start by multiplying each side of the equation by 5 to get $2x-12=6\cdot 5=30$. So $2x=30+12=42$. Finally, $x=\frac{42}{2}=\mathbf{21}$.

Example 11.7: For what value of x does $\frac{x+20}{3x}=2$?

Solution: We start by multiplying each side of the equation by $3x$ to get $x+20=2\cdot 3x=6x$. So $20=5x$, and $x=\frac{20}{5}=\mathbf{4}$.

Example 11.8: If $\frac{2x+3}{x-2}=\frac{3}{4}$, then what is the value of x ?

Solution: We start by multiplying each side of the equation by $4(x-2)$ to get $4(2x+3)=3(x-2)$. We now distribute the 4 on the left and the 3 on the right to get $8x+12=3x-6$. So $5x=-18$, and $x=-\mathbf{\frac{18}{5}}$.

Notes: (1) The least common multiple of the two denominators or least common denominator (LCD) is $4(x-2)$.

In this particular example, the LCD is the same as the product of the denominators.

(2) Multiplying each side of the given equation by the LCD is equivalent to *cross multiplying*. In other words,

$\frac{a}{b} = \frac{c}{d}$ is equivalent to the equation $ad = bc$.

(3) Since there is a variable in a denominator of the original equation, we should check for extraneous solutions at the end. Since $x = -\frac{18}{5}$ does not make any denominator zero, it is *not* extraneous. That is , $x = -\frac{18}{5}$ is actually a solution.

Example 11.9: Solve for x: $6x - (3 - x) = 18 - (3 - x)$.

Solution: We start by distributing the "subtraction symbols" to get $6x - 3 + x = 18 - 3 + x$. So $6x = 18$, and therefore $x = \frac{18}{6} = \mathbf{3}$.

Example 11.10: Find x if $2x + 4 = x + 8$.

Solution: $2x + 4 = x + 8 \Leftrightarrow x = \mathbf{4}$.

Example 11.11: If $4(x + 3) = 2(x - 3)$, what is the value of x ?

Solution:

$4(x + 3) = 2(x - 3) \Leftrightarrow 4x + 12 = 2x - 6 \Leftrightarrow 2x = -18 \Leftrightarrow x = -\frac{18}{2} = \mathbf{9}$.

LINEAR SYSTEMS OF TWO EQUATIONS IN TWO UNKNOWNS

Here is an example of a linear system of two equations in two unknowns:

$$2x = 7 - 3y$$
$$5y = 5 - 3x$$

Let's go over two ways to solve this system for x.

Method 1 – elimination: We begin by making sure that the two equations are "lined up" properly. We do this by adding $3y$ to each side of the first equation, and adding $3x$ to each side of the second equation.

$$2x + 3y = 7$$
$$3x + 5y = 5$$

We will now multiply each side of the first equation by 5, and each side of the second equation by -3.

$$5(2x + 3y) = (7)(5)$$
$$-3(3x + 5y) = (5)(-3)$$

Do not forget to distribute correctly on the left. Add the two equations.

$$10x + 15y = 35$$
$$\underline{-9x - 15y = -15}$$
$$x \qquad = \mathbf{20}$$

Remarks: (1) We chose to use 5 and -3 because multiplying by these numbers makes the y column "match up" so that when we add the two equations in the next step the y term vanishes. We could have also used -5 and 3.

(2) If we wanted to find y instead of x we would multiply the two equations by 3 and -2 (or -3 and 2). In general, if you are only looking for one variable, try to eliminate the one you are **not** looking for.

(3) We chose to multiply by a negative number so that we could add the equations instead of subtracting them. We could have also multiplied the first equation by 5, the second by 3, and subtracted the two equations, but a computational error is more likely to occur this way.

Method 2 – substitution: We solve the second equation for y and substitute into the first equation.

$5y = 5 - 3x$ implies $y = \frac{5-3x}{5} = \frac{5}{5} - \frac{3x}{5} = 1 - \frac{3x}{5}$. So now using the first equation we have

$$2x = 7 - 3y = 7 - 3(1 - \frac{3x}{5}) = 7 - 3 + \frac{9x}{5} = 4 + \frac{9x}{5}.$$

Multiply each side of this equation by 5 to get rid of the denominator on the right. So we have $10x = 20 + 9x$, and therefore $x = \mathbf{20}$.

Remark: If we wanted to find y instead of x we would solve the first equation for x and substitute into the second equation.

Example 11.12: Solve the system of equations by the elimination method: $x + y = 11, x - y = 5$.

Solution: We add the equations to get $2x = 16$. So $x = \frac{16}{2} = 8$. Using the first equation to find y, we have $y = 11 - x = 11 - 8 = 3$.

So the system of equations has the unique solution $(\mathbf{8}, \mathbf{3})$.

Example 11.13: Find the solution set of the system $\left.\begin{matrix} x + y = 12 \\ x - y = 8 \end{matrix}\right\}$ by the elimination method.

Solution: We add the two equations to get $2x = 20$, and so $x = \frac{20}{2} = 10$. It follows that $y = 2$, and so the unique solution is $(10,2)$.

Example 11.14: Solve the system of equations $\left.\begin{matrix} 3x - 2y = 7 \\ 6x + 3y = 21 \end{matrix}\right\}$ by the elimination method.

Solution: We first multiply each side of the first equation by -2.

$$-2(3x - 2y) = (7)(-2)$$
$$6x + 3y = 21$$

Do not forget to distribute correctly on the left. Add the two equations.

$$\begin{array}{c} -6x + 4y = -14 \\ \underline{6x + 3y = 21} \\ 7y = 7 \end{array}$$

So $y = \frac{7}{7} = 1$. Using the first equation, we have

$$3x = 7 + 2y = 7 + 2 \cdot 1 = 7 + 2 = 9,$$

and so $x = \frac{9}{3} = 3$. So the system of equations has the unique solution $(\mathbf{3}, \mathbf{1})$.

Example 11.15: $\left.\begin{array}{l} 4x + 3y = 25 \\ 5x + 2y = 26 \end{array}\right\}$ Find the solution set of the given system.

Solution: We first multiply each side of the first equation by -2, and each side of the second equation by 3.

$$(-2)(4x + 3y) = 25(-2)$$
$$3(5x + 2y) = 26 \cdot 3$$

Do not forget to distribute correctly on the left. Add the two equations.

$$\begin{array}{c} -8x - 6y = -50 \\ \underline{15x + 6y = 78} \\ 7x = 28 \end{array}$$

So $x = \frac{28}{7} = 4$. Using the first equation to find y, we have

$$3y = 25 - 4x = 25 - 4 \cdot 4 = 25 - 16 = 9.$$

So $y = \frac{9}{3} = 3$, and the system of equations has the unique solution $(\mathbf{4}, \mathbf{3})$.

Example 11.16: $\left.\begin{array}{l} x - 2y = 1 \\ 3x + 4y = 23 \end{array}\right\}$ Solve the given system by using the substitution method.

Solution: We solve the first equation for x and substitute into the second equation.

$x - 2y = 1$ implies$x = 1 + 2y$. So now using the second equation we have

$$4y = 23 - 3x = 23 - 3(1 + 2y) = 23 - 3 - 6y = 20 - 6y.$$

Adding $6y$ to each side of his last equation yields $10y = 20$.

So $y = \frac{20}{10} = 2$. We now substitute $y = 2$ into either of the original equations. Let's use the first equation. Then we have

$$x = 1 + 2y = 1 + 2 \cdot 2 = 1 + 4 = 5$$

So the system of equations has the unique solution $(\mathbf{5}, \mathbf{2})$.

12. COUNTING

MAKING A LIST

Sometimes the easiest way to count the number of possibilities is to simply list them all. When doing this it is important that you have a systematic way of forming your list. This will reduce the likelihood of missing something, or listing something twice.

Example 12.1: A menu lists 2 meals and 3 drinks. How many different meal drink combinations are possible from this menu?

Solution: Let's make up 2 meals and 3 drinks, say chicken and fish for the meals, and water, soda, and juice for the drinks. Now let's list all meal-drink combinations. To save time we will abbreviate each meal and drink by using its first letter.

cw cs cj fw fs fj

We see that there are **6** meal-drink combinations.

Note: We can actually solve this problem very quickly by using the *counting principle*: (2)(3) = 6.

THE COUNTING PRINCIPLE

The **counting principle** (or **multiplication principle)** says that if one event is followed by a second independent event, the number of possibilities is multiplied.

For example, the 2 events in example 12.1 are "choosing a meal," and "choosing a drink."

More generally this principle says that if $E_1, E_2, \ldots, E_n$ are n independent events with $m_1, m_2, \ldots, m_n$ possibilities, respectively, then event E_1, followed by event E_2, followed by event $E_3, \ldots$, followed by event E_n has $m_1 \cdot m_2 \cdots m_n$ possibilities.

Example 12.2: A menu lists 2 appetizers, 5 meals, 4 drinks, and 3 desserts. A dinner consists of 1 of each of these 4 items. How many different dinners are possible from this menu?

Solution: We use the counting principle to get (2)(5)(4)(3) = **120**.

Example 12.3: How many integers between 9 and 300 have the tens digit equal to 2, 3, or 4 and the units digit (ones digit) equal to 5 or 6?

Solution 1: We use the counting principle:

There are 2 possibilities for the ones digit (5 or 6). There are 3 possibilities for the tens digit (2, 3, or 4). There are 3 possibilities for the hundreds digit (0, 1, or 2). The counting principle says that we multiply the possibilities to get (2)(3)(3) = **18**.

Solution 2: Let's simply list all the numbers in increasing order:

25 26 35 36 45 46 125 126 135 136 145 146 225 226 235 236 245 246

We see that the answer is **18**.

THE ADDITION PRINCIPLE

The **addition principle** says that if E_1 and E_2 have nothing in common, E_1 has m_1 possibilities and E_2 has m_2 possibilities, then E_1 or E_2 has $m_1 + m_2$ possibilities.

Example 12.4: In how many different ways could one math book or one chemistry book be selected from 4 math and 6 chemistry books each with different content?

Solution: By the addition principle the answer is $4 + 6 = \mathbf{10}$.

PERMUTATIONS

Recall: The **factorial** of a positive integer n, written $n!$, is the product of all positive integers less than or equal to n: $\mathbf{n! = 1 \cdot 2 \cdot 3 \cdots n}$

0! is defined to be 1, so that n! is defined for all nonnegative integers n.

A **permutation** is just an arrangement of elements from a set. The number of permutations of n things taken r at a time is

$${}_nP_r = \frac{n!}{(n-r)!}$$

Notation: The permutation ${}_n\mathrm{P_r}$ can also be written $\mathrm{P(n, r)}$.

Example 12.5: The number of permutations of {1, 2, 3} taken 2 at a time is

$$_3P_2 = \frac{3!}{1!} = 6$$

These permutations are 12, 21, 13, 31, 23, and 32.

Note: You can do this computation very quickly on your TI-84 calculator. To compute $_3P_2$, type 3 into your calculator, then in the **Math** menu scroll over to **Prb** and select **nPr** (or press **2**). Then type 2 and press **Enter**. You will get an answer of 6.

Example 12.6: Four different books are to be stacked in a pile. In how many different orders can the books be placed on the stack?

Solution: There are 4 books, and we are arranging all 4 of them. Therefore the number of arrangements is $_4P_4 = 4! = (1)(2)(3)(4) =$ **24**.

Example 12.7: Three light bulbs are placed into three different lamps. How many different arrangements are possible for three light bulbs of different colors – one red, one green, and one yellow?

Solution: There are 3 light bulbs, and we are arranging all 3 of them. So the number of arrangements is $_3P_3 = 3! = 1 \cdot 2 \cdot 3 =$ **6**.

COMBINATIONS

A **combination** is just a subset containing a specific number of the elements of a particular set. The number of combinations of n things taken r at a time is

$$_nC_r = \frac{n!}{r!(n-r)!}.$$

Notation: The combination $_nC_r$ can also be written $C(n,r)$ or $\binom{n}{r}$.

Example 12.8: The number of combinations of {1, 2, 3} taken 2 at a time is

$$_3C_2 = \frac{3!}{2!1!} = 3$$

These combinations are 12, 13, and 23.

Notes: (1) 21 is the **same** combination as 12 (but 12 and 21 are **different** permutations).

(2) You can do this computation very quickly on your TI-84 graphing calculator. To compute $_3C_2$, type 3 into your calculator, then in the Math menu scroll over to **Prb** and select **nCr** (or press **3**). Then type 2 and press **Enter**. You will get an answer of **3**.

Combination Rules:

1) ${}_{n}C_{0} = 1$
2) ${}_{n}C_{1} = n$
3) ${}_{n}C_{n} = 1$
4) If ${}_{n}C_{x} = {}_{n}C_{y}$, then $x = y$ or $x + y = n$
5) ${}_{n}C_{0} + {}_{n}C_{1} + {}_{n}C_{2} + \cdots + {}_{n}C_{n} = 2^{n}$

DIFFERENCE BETWEEN PERMUTATIONS AND COMBINATIONS

"Permutation" is just a fancy word for arrangement. Whenever you are counting the number of ways to arrange things you can use the permutation formula (or when possible the **nPr** feature in your calculator). So if you want to "order" things, "stack" things, "rearrange" things, choose specific tasks for people, etc, you would use permutations.

"Combinations" are used when you want to group things together without putting them in any specific order. This happens when you split a group of objects into two piles, or form a committee of people, for example.

So the numbers 12 and 21 are different permutations, but the same combination.

Example 12.9: How many committees of 4 people can be formed from a group of 9?

Solution: The order in which we choose the 4 people does not matter. Therefore this is the combination ${}_{9}C_{4}$ = **126.**

Example 12.10: A chemist is testing 5 different liquids. For each test, the chemist chooses 3 of the liquids and mixes them together. What is the least number of tests that must be done so that every possible combination of liquids is tested?

Solution: We are counting the number of ways to choose 3 of the 5 liquids. This is ${}_{5}C_{3}$ = **10.**

Note: We can also solve this problem by simply listing all of the possibilities. But be careful! This is just a bit tricky.

In the following list a * means we are choosing that liquid, and an O means we are not:

***OO	*OO**
**O*O	O***O
**OO*	O**O*
*O**O	O*O**
*O*O*	OO***

We see that there are 10 combinations.

Warning: Just because the word "combination" appears in a counting problem, it does not mean that the problem should be solved by using a combination in the mathematical sense.

Example 12.11: A wall is to be painted one color with a stripe of a different color running through the middle. If 7 different colors are available, how many color combinations are possible?

Solution: There are ${}_7P_2$ = **42** ways to choose 2 colors from 7, and place them in a specific order.

Important note: Don't let the word "combinations" in the problem itself trick you. This is **not** a combination in the mathematical sense. If you paint the wall red and the stripe blue, then this is a **different** choice from painting the wall blue and the stripe red.

MIXED COUNTING PROBLEMS

Example 12.12: In how many different ways could one math and one physics book be selected among 3 math and 4 physics books, all different?

Solution: We use the counting principle to get (3)(4) = **12**.

Example 12.13: In how many different ways can 8 people sit in an armchair that fits two people? Assume that 2 people must sit in the chair.

Solution: This is the permutation $P(8,2) = 8 \cdot 7 = \mathbf{56}$.

Example 12.14: Evaluate $\boldsymbol{C(16,2)}$.

Solution: $C(16,2) = \frac{16!}{(16-2)!2!} = \frac{16\cdot15\cdot14!}{14!2!} = \frac{16\cdot15}{2} = 8 \cdot 15 = \mathbf{120}$.

Example 12.15: In how many different ways can a team of 2 students be chosen from 8 students?

Solution: This is the combination $C(8,2) = \frac{8!}{6!2!} = \mathbf{28}$.

Example 12.16: How many three digit positive integers with all digits distinct can we form using the digits 0, 1, 2, 3, 4, 5 ?

Solution: The leftmost digit must be an integer between 1 and 5, inclusive. So there are 5 choices for the leftmost digit.

For the middle digit we must choose an integer between 0 and 5 that is not equal to the leftmost digit. So we again have 5 choices.

For the rightmost digit we must choose an integer between 0 and 5 that is not equal to the other two digits. So we have 4 choices.

By the counting principle, the total number of three digit positive integers with all digits distinct is $5 \cdot 5 \cdot 4 = \mathbf{100}$.

Note: The leftmost digit cannot be 0 because then we would wind up with a two digit positive integer instead of a three digit positive integer.

Example 12.17: In how many different ways can one man and one woman be selected from five men and six women?

Solution: 1 man can be selected in 5 different ways from 5 men, and 1 woman can be selected in 6 different ways from 6 women. We then use the counting principle to get $5 \cdot 6 = \mathbf{30}$.

Example 12.18: Find the number of 3 element combinations from a set with 9 elements.

Solution: $C(9,3) = \frac{9!}{(9-3)!3!} = \frac{9 \cdot 8 \cdot 7 \cdot 6!}{6!3!} = \frac{9 \cdot 8 \cdot 7}{3 \cdot 2 \cdot 1} = 3 \cdot 4 \cdot 7 = \mathbf{84}$.

Example 12.19: In how many different ways can a team of 3 students be selected from 12 students?

Solution: $C(12,3) = \frac{12!}{(12-3)!3!} = \frac{12 \cdot 11 \cdot 10 \cdot 9!}{9!3!} = \frac{12 \cdot 11 \cdot 10}{3 \cdot 2 \cdot 1} = 2 \cdot 11 \cdot 10 = \mathbf{220}$.

Example 12.20: Let $A = \{0,1,2,3,4,5,6,7\}$. How many four digit positive integers can be formed using only digits from A ?

Solution: The leftmost digit must be an integer between 1 and 7, inclusive. So there are 7 choices for the leftmost digit.

Each of the other 3 digits must be an integer between 0 and 7. So we have 8 choices for each of the remaining 3 digits.

By the counting principle, the total number of four digit positive integers with all digits from A is $7 \cdot 8 \cdot 8 \cdot 8 = \mathbf{3584}$.

Note: The leftmost digit cannot be 0 because then we would wind up with a three digit positive integer instead of a four digit positive integer.

Example 12.21: Evaluate ${}_6C_3 - {}_6C_4$.

Solution: ${}_6C_3 - {}_6C_4 = \frac{6!}{(6-3)!3!} - \frac{6!}{(6-4)!4!} = \frac{6\cdot5\cdot4\cdot3!}{3!3!} - \frac{6\cdot5\cdot4!}{2!4!} = 20 - 15 = \mathbf{5}$.

Example 12.22: Ahmet has 9 pens of different colors. In how many ways can he give 3 of these pens to his brother?

Solution: $C(9,3) = \frac{9!}{(9-3)!3!} = \frac{9\cdot8\cdot7\cdot6!}{6!3!} = \frac{9\cdot8\cdot7}{3\cdot2\cdot1} = 3\cdot4\cdot7 = \mathbf{84}$.

Example 12.23: How many three digit even positive integers with all digits distinct can we form using the digits 1, 2, 3, 4, 5, 6, 7, 8 ?

Solution: The rightmost digit must be 2, 4, 6, or 8. So there are 4 choices for the rightmost digit.

For the middle digit we must choose an integer between 1 and 8, inclusive, that is not equal to the rightmost digit. So we have 7 choices.

For the leftmost digit we must choose an integer between 1 and 8 that is not equal to the other two digits. So we have 6 choices.

By the counting principle, the total number of three digit positive integers with all digits distinct is $4\cdot7\cdot6 = \mathbf{168}$.

Notes: (1) An integer is even when its rightmost digit is even. That is, an even integer ends in 0, 2, 4, 6, or 8.

In this problem 0 is not allowed.

(2) When using the counting principle it is usually easiest to deal with "special cases" first. In this case, the rightmost digit is the special case because there is a restriction on which digits can be used.

Example 12.24: In how many different ways can 6 books be lined up on a bookshelf?

Solution: We are simply arranging the books here so that this is a straightforward permutation: $P(6,6) = 6! = \mathbf{720}$.

Example 12.25: In how many different ways can 7 people sit around a round table?

Solution: We begin placing the first person arbitrarily. Moving around clockwise, there are 6 choices for the next seat, 5 for the one after that, and so on. So the answer is $6! = \mathbf{720}$.

Note: This is an example of a **cyclic permutation**. The number of ways to arrange n items in a circle is $(n-1)!$.

Example 12.26: Among 10 students, 5 will be sent to Kansas, and 5 will be sent to Arkansas. In how many different ways can the students be selected for these two trips?

Solution:

$$C(10,5) \cdot C(5,5) = \frac{10!}{(10-5)!5!} \cdot \frac{5!}{(5-5)!5!} = \frac{10 \cdot 9 \cdot 8 \cdot 7 \cdot 6}{5 \cdot 4 \cdot 3 \cdot 2 \cdot 1} \cdot 1 = 2 \cdot 3 \cdot 2 \cdot 7 \cdot 3 = \mathbf{252}.$$

Notes: (1) We first choose 5 of the 10 students to go to Kansas. There are $C(10,5)$ ways to do this.

We then choose 5 of the remaining 5 student to go to Arkansas. There are $C(5,5)$ ways to do this.

By the counting principle we multiply the possibilities: $C(10,5) \cdot C(5,5)$.

(2) Once we choose the 5 students to go to Kansas, the other 5 must go to Arkansas. So we actually do not need to multiply by $C(5,5)$. In fact, the number of ways to choose 5 people from 5 people is just 1, ie. $C(5,5) = 1$.

Example 12.27: There are 6 plates of stuffed grape leaves, 5 plates of rice, and 7 plates of baklava. In how many different ways can we select 1 plate of each?

Solution: We use the counting principle to see that the number of selections is $6 \cdot 5 \cdot 7 = \mathbf{210}$.

Example 12.28: Compute $C(6,0)$.

Solution: $C(6,0) = \frac{6!}{(6-0)!0!} = \frac{6!}{6!0!} = \frac{1}{0!} = \frac{1}{1} = \mathbf{1}$.

Recall: 0!=1.

Example 12.29: Among 12 students, 6 will be sent to Kansas, 4 will be sent to Arkansas, and 2 will be sent to Oklahoma. In how many different ways can the students be selected for these three trips?

Solution: $C(12,6) \cdot C(6,4) \cdot C(2,2) = \frac{12!}{(12-6)!6!} \cdot \frac{6!}{(6-4)!4!} \cdot 1 = \frac{12 \cdot 11 \cdot 10 \cdot 9 \cdot 8 \cdot 7}{6 \cdot 5 \cdot 4 \cdot 3 \cdot 2 \cdot 1} \cdot \frac{6 \cdot 5}{2 \cdot 1}$

$$= 2 \cdot 11 \cdot 2 \cdot 3 \cdot 7 \cdot 3 \cdot 5 = \mathbf{13,860}.$$

Example 12.30: A committee of four people and an additional representative will be selected from 8 people. In how many different ways can these people be selected?

Solution: $C(8,4) \cdot C(4,1) = \frac{8!}{(8-4)!4!} \cdot 4 = \frac{8 \cdot 7 \cdot 6 \cdot 5}{4 \cdot 3 \cdot 2 \cdot 1} \cdot 4 = 2 \cdot 7 \cdot 5 \cdot 4 = \mathbf{280}$.

Example 12.31: A set A has n elements. The number of combinations taken 6 at a time from set A is equal to the number of combinations taken 7 at a time. What is the value of n ?

Solution: We are given $C(n, 6) = C(n, 7)$. It follows that $n = 6 + 7 = \mathbf{13}$.

Note: See combination rule 4 above.

Example 12.32: A teacher gives 15 questions on an exam. Students must answer the first 5 questions, and then choose 8 of the remaining questions to answer. In how many ways can a student taking the exam choose the questions he or she answers?

Solution: After answering the first 5 questions, the student must choose 8 of the remaining 15 – 5 = 10 questions. $C(10,8) = \frac{10\cdot 9}{2\cdot 1} = \mathbf{45}$.

Example 12.33: Eight parallel lines, all with the same positive slope, intersect 7 other parallel lines, all with the same negative slope. How many parallelograms are formed from these lines?

Solution: $C(8,2) \cdot C(7,2) = \frac{8\cdot 7}{2\cdot 1} \cdot \frac{7\cdot 6}{2\cdot 1} = 4 \cdot 7 \cdot 7 \cdot 3 = \mathbf{588}$.

Example 12.34: Ahmet would like to invite seven of his ten friends to a party. Four of the seven friends to be invited have already been determined. How many choices does Ahmet have?

Solution: $C(10 - 4, 7 - 4) = C(6,3) = \frac{6\cdot 5\cdot 4}{3\cdot 2\cdot 1} = 5 \cdot 4 = \mathbf{20}$.

13. SETS AND PROBABILITY

SETS

The objects in a set are called **elements**.

If b is an element of the set A, we write $b \in A$, and if b is not an element of A, we write $b \notin A$.

We usually assume that we have a u**niversal set**, a set which contains all other sets under discussion. Here is a picture of a universal E containing another set A.

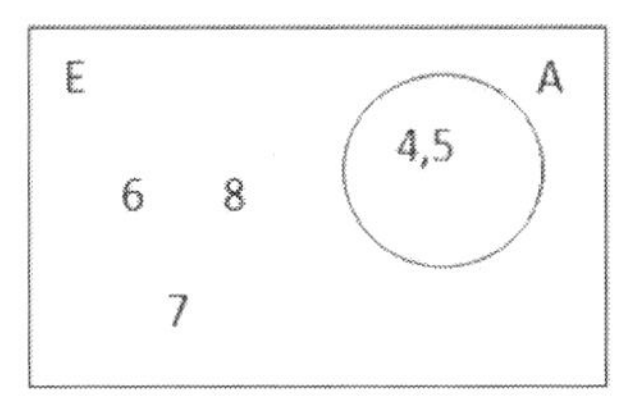

In the above figure the universal set is $E = \{4,5,6,7,8\}$, and $A = \{4,5\}$.

REPRESENTING SETS

There are several ways to describe a set.

Listing method: The elements in the set are listed inside curly braces { }, and a comma is written between the elements.

Example 13.1: Express the set P of the prime numbers between 1 and 10 using the listing method.

Solution: $P = \{2,3,5,7\}$

Venn diagram: The elements of the set are placed in circles. Distinct sets generally have circles which intersect unless it is clear that the sets have nothing in common.

Example 13.2: Express the sets $A = \{x, y, z, m\}$ and $B = \{x, z, n\}$ using a Venn diagram.

Solution:

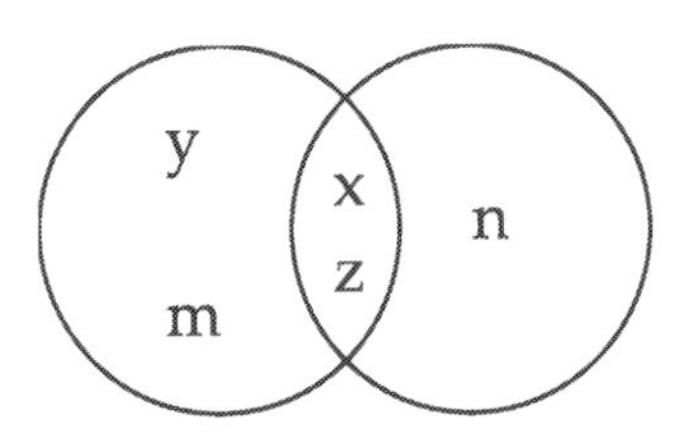

Note: The question itself used the listing method to describe the two sets.

Set builder notation: We write $\{x|P(x)\}$. This is read "the set of x such that $P(x)$." Here $P(x)$ is some property describing what it means to be in the set.

Example 13.3: Express the set $\{2,4,6,8,10,12\}$ in set builder notation.

Solution: $\{n|n$ is an even positive integer less than or equal to $12\}$.

Note: A description of a set using set builder notation is not unique. There are generally many equivalent ways to describe a set.

The **cardinality** of a finite set A, written $|A|$ is the number of elements it has. So if A has n elements, we write $|A| = n$.

Example 13.4: If $A = \{a, b, c, d, e\}$, then $|A| = 5$.

We say that a set A is a **subset** of a set B if every element of A is an element of B. In this case we write $A \subset B$. If $A \neq B$, then we may also say that A is a **proper subset** of B.

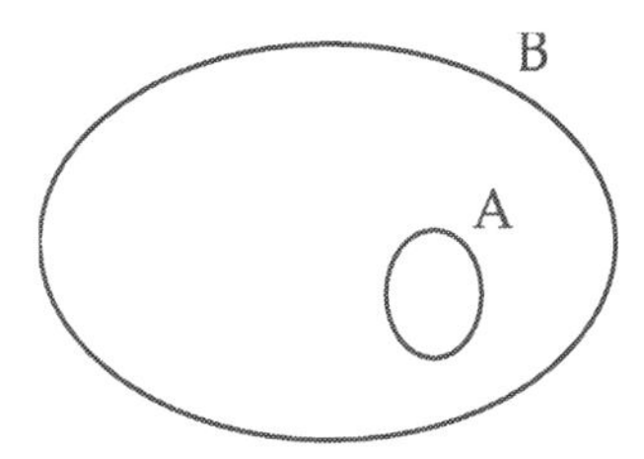

Example 13.5: Let $A = \{1,2,3,4,5\}$ and $B = \{1,3,5\}$. Draw a Venn diagram showing that $B \subset A$.

Solution:

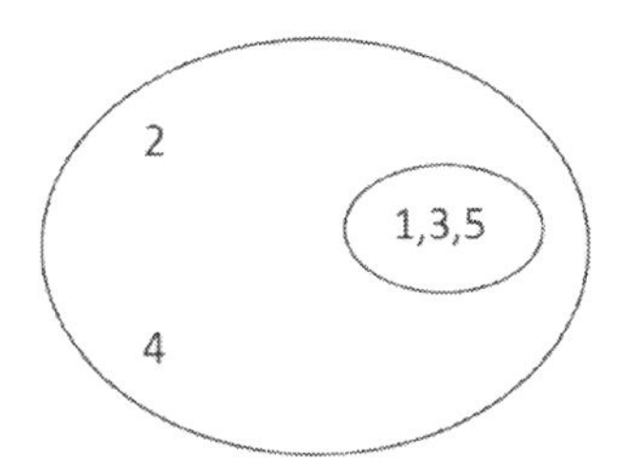

The **empty set** is the unique set with no elements. We use the symbol $\emptyset$ to represent the empty set.

Note that $\emptyset$ is a subset of every set. In particular, $\emptyset \subset \emptyset$.

A set with n elements has 2^n subsets.

For example, the empty set has $2^0 = 1$ subset. Indeed, the only subset of $\emptyset$ is $\emptyset$ itself.

A set with 1 element has $2^1 = 2$ subsets. For example, the subsets of $A = \{1\}$ are $\emptyset$ and A.

A set with 2 elements has $2^2 = 4$ subsets. For example, the subsets of $B = \{1,2\}$ are $\emptyset$, $\{1\}$, $\{2\}$, and B.

A set with 3 elements has $2^3 = 8$ subsets. For example, the subsets of $C = \{1,2,3\}$ are $\emptyset$, $\{1\}$, $\{2\}$, $\{3\}$, $\{1,2\}$, $\{1,3\}$, $\{2,3\}$, and C.

Example 13.6: If $A = \{a, b, c, d, e\}$, find the number of the subsets of A.

Solution: Since $|A| = 5$, the number of subsets of A is $2^5 = \mathbf{32}$.

Example 13.7: If $X = \{1,2,3,4,5,6\}$, find the number of the subsets of X.

Solution: Since $|X| = 6$, the number of subsets of X is $2^6 = \mathbf{64}$.

Example 13.8: The number of subsets of a set Y is 512. What is $|Y|$?

Solution: If $|Y| = n$, then $2^n = 512 = 2^9$. So $n = \mathbf{9}$.

Example 13.9: number of proper subsets of set K is 63. What is $|K|$?

Solution: The number of subsets of K is $63+1=64$. If $|K|=n$, then $2^n=64=2^6$. So $n=\mathbf{6}$.

Example 13.10: If $A=\{7,8,9\}$ and $B=\{5,6,7,8,9,10,11\}$, draw a Venn diagram showing that B contains A.

Solution:

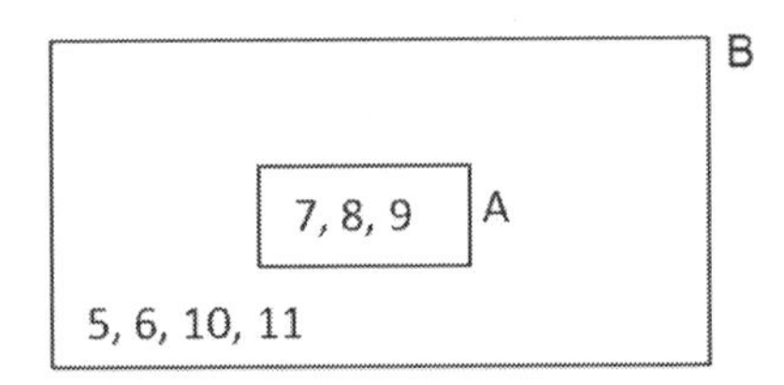

$B \supset A \rightarrow$ set B contains set A. This is equivalent to $A \subset B$ (A a subset of B).

Example 13.11: How many 1 element subsets does a 9 element set have?

Solution: Let $A=\{1,2,3,4,5,6,7,8,9\}$. The 1 element subsets of A are $\{1\},\{2\},\{3\},\{4\},\{5\},\{6\},\{7\},\{8\},\{9\}$. We see that the number of such subsets is **9**.

Example 13.12: What is the number of subsets of $A=\{x \in Z \mid -3 \leq x \leq 3\}$

Solution: $A=\{-3,-2,-1,0,1,2,3\}$. So $|A|=7$, and so the number of subsets of A is $2^7=\mathbf{128}$.

Example 13.13: Find the number of proper subsets of the set $A=\{1,2,3,4,5\}$.

Solution: $|A|=5$. So A has $2^5=32$ subsets. Therefore the number of proper subsets of A is $32-1=\mathbf{31}$.

SET OPERATIONS

The **union** of A and B, written $A \cup B$ is the set which contains elements that are in A or B or both.

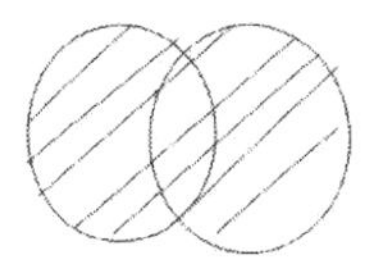 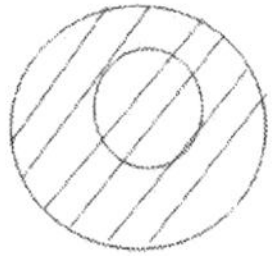 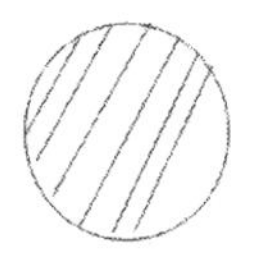 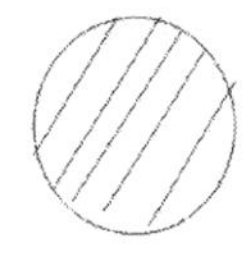

The three figures above show Venn diagrams of $A \cup B$ in different situations. The leftmost figure is the most general way to draw $A \cup B$. If we know that $A \subset B$, then $A \cup B$ can be drawn like the middle figure.

And if A and B have no elements in common, then we can draw $A \cup B$ like the rightmost figure.

Properties of unions

1) $A \cup A = A$
2) $A \cup B = B \cup A$
3) $A \cup E = E$ The universal set contains every set.
4) $A \cup \emptyset = A$ Every set contains the empty set.
5) If $A \subset B$, then $A \cup B = B$.

Example 13.14: If $A = \{2,4,6,8,10,12\}$ and $B = \{10,12,14,16\}$, find $A \cup B$.

Solution: $A \cup B = \{2,4,6,8,10,12,14,16\}$

Example 13.15:

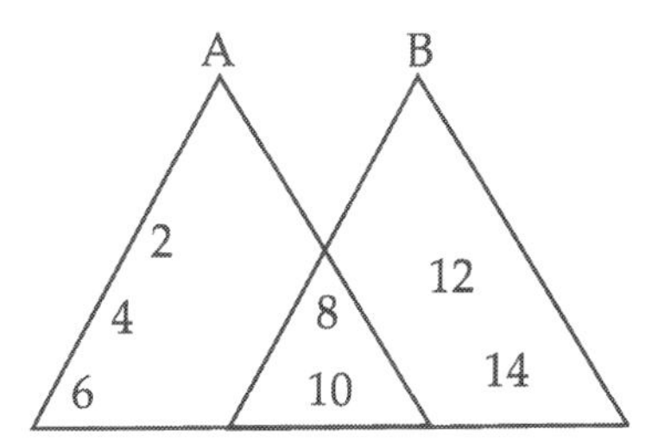

Use the Venn diagram above to compute $A \cup B$.

Solution: $A \cup B = \{2,4,6,8,10,12,14\}$

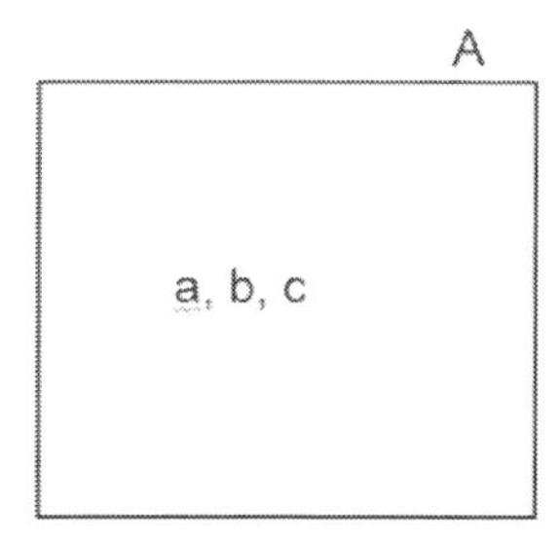

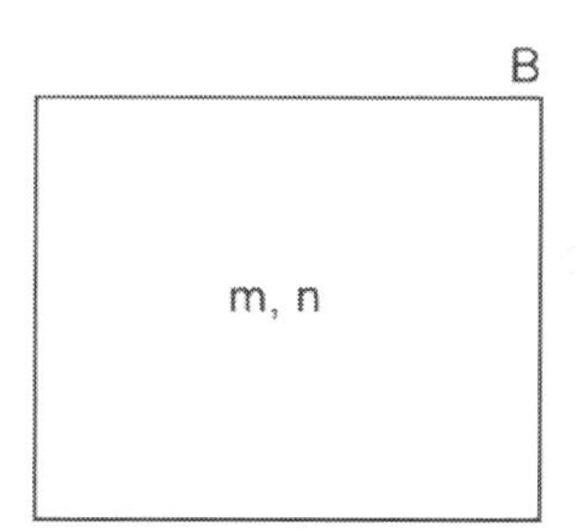

Example: Use the Venn diagram above to find $A \cup B$?

Solution: $A \cup B = \{a, b, c, m, n\}$

Example 13.16: Given that $|A| = 6$ and $|B| = 8$, what is the maximum value of $|A \cup B|$?

Solution: $A \cup B$ will be largest if A and B have no elements in common. In this case $|A \cup B| = 6 + 8 = \mathbf{14}$.

Note: It is *not* true that $|A \cup B|$ must be 14. For example, if $A \subset B$, then $|A \cup B| = |B| = 8$.

The **intersection** of A and B, written $A \cap B$ is the set which contains all elements common to A and B.

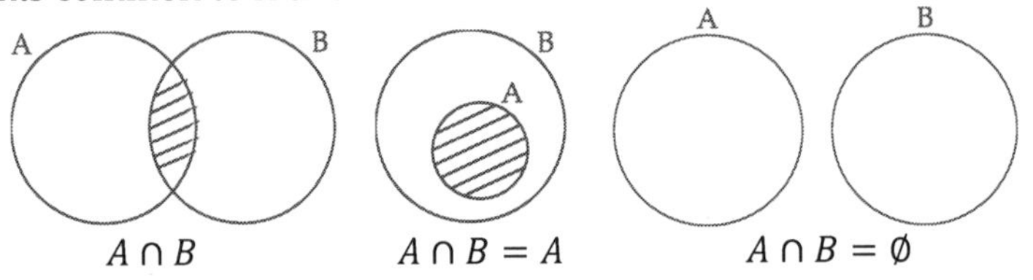

The three figures above show Venn diagrams of $A \cap B$ indifferent situations. The leftmost figure is the most general way to draw $A \cap B$. If we know that $A \subset B$, then $A \cap B$ can be drawn like the middle figure. And if A and B have no elements in common, then we can draw $A \cap B$ like the rightmost figure.

Properties of intersections:

1) $A \cap A = A$
2) $A \cap B = B \cap A$
3) $A \cap E = A$
4) $A \cap \emptyset = \emptyset$
5) If $A \subset B$, then $A \cap B = A$
6) $A \cup (B \cap C) = (A \cup B) \cap (A \cup C)$
7) $A \cap (B \cup C) = (A \cap B) \cup (A \cap C)$
8) **Addition Principle:** $|A \cup B| = |A| + |B| - |A \cap B|$

Example 13.17: If $A = \{2,4,6,8,10\}$ and $B = \{6,8,10,12,14\}$, find $A \cap B$.

Solution: $A \cap B = \{6,8,10\}$.

The **difference** $A \setminus B$ is the set which contains all elements of A that are not in B. $A \setminus B = \{x | x \in A, x \notin B\}$

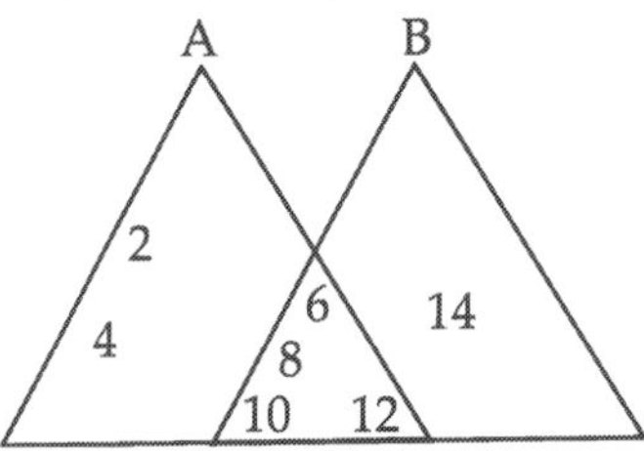

Example 13.18: Find $A \setminus B$, $B \setminus A$ and $A \cap B$ in the figure above.

Solution: $A \setminus B = \{2,4\}$, $B \setminus A = \{14\}$, and $A \cap B = \{6,8,10,12\}$.

Example 13.19: Suppose that $A = \{2,4,6\}$, $A \cup B = \{2,4,6,8,10\}$, and $A \setminus B = \{2,4\}$. Find B.

Solution: Let's draw a Venn diagram:

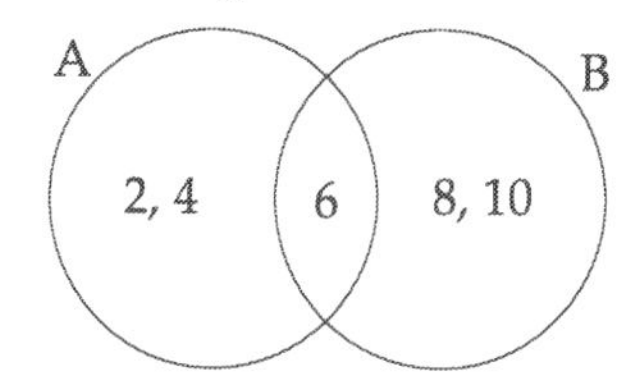

We see that $B = \{6,8,10\}$.

Example 13.20: Suppose that $A = \{1,2,3,4\}$ and $B = \{5,6,7\}$. Find $A \cup B$.

Solution: Let's draw a Venn diagram:

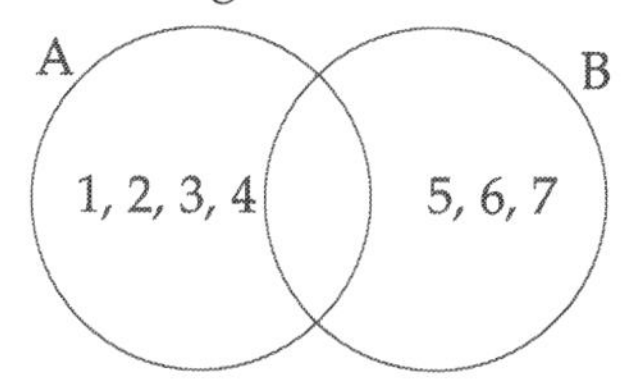

We see that $A \cup B = \{1,2,3,4,5,6,7\}$.

Example 13.21: Let A and B be sets with $|A| = 5 \cdot |B|$, $|A \cap B| = 14$, and $|A \cup B| = 34$. Find $|B|$.

Solution: We have $|A \cup B| = |A| + |B| - |A \cap B|$. So $34 = 5|B| + |B| - 14$. Therefore $6|B| = 34 + 14 = 48$, and so $|B| = \frac{48}{6} = \mathbf{8}$.

Example 13.22: If $A \cup B = \{10,20,30,40,50\}$ and $A \cup C = \{20,40,60,70,80\}$, find $|A \cup (B \cap C)|$.

Solution: $A \cup (B \cap C) = (A \cup B) \cap (A \cup C) = \{20,40\}$. So $|A \cup (B \cap C)| = \mathbf{2}$.

Example 13.23: If $A = \{a, b, c, d, e\}$ and $B = \{a, e, f\}$, what is $A \cup B$?

Solution: $A \cup B = \{a, b, c, d, e, f\}$

Example 13.24: If $A = \{2,4,6,8,10\}$ and $B = \{2,4,9,10\}$, what is $A \cap B$?

Solution: $A \cap B = \{2,4,10\}$

Example 13.25: Let $A = \{n | n \text{ is a potive integer less than } 10\}$ and let $B = \{n | n \text{ is a prime number less than } 8\}$. Find $A \cap B$.

Solution: $A = \{1,2,3,4,5,6,7,8,9\}$ and $B = \{2,3,5,7\}$. So $A \cap B = \{2,3,5,7\}$

Example 13.26: Let $A = \{2,4,6,8\}$ and $B = \{2,10,12\}$. Find $A \setminus B$ and $B \setminus A$.

Solution: $A \setminus B = \{4,6,8\}$ and $B \setminus A = \{10,12\}$.

Example 13.27: Display the set of the prime numbers less than 20 with the listing method and using set-builder notation.

Solution: The set of primes less than 20 is $P = \{2,3,5,7,11,13,17\}$ (listing method). Using set builder notation we have

$$P = \{p \in Z^{+} \mid p < 20 \text{ and } p \text{ is prime}\}.$$

Example 13.28:

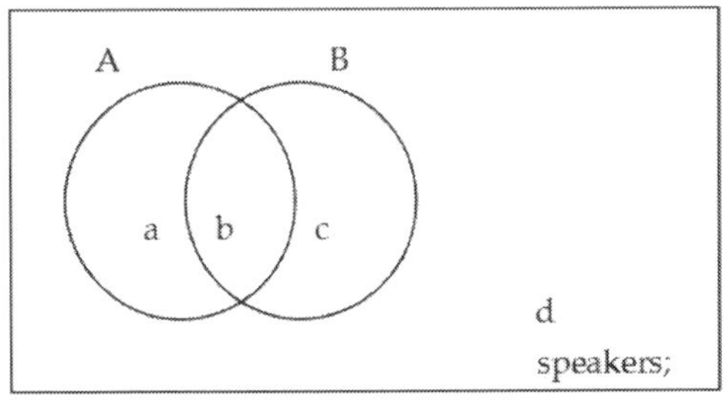

A: set of english speakers

B: set of german speakers

group of (a+b+c+d) people that

includes English and german

- Number of English speakers only: a
- Number of German speakers only: c
- Number of German-English speakers: b
- Number of English speakers: $a + b$
- Number of German speakers: $b + c$
- Number of people who speak English or German: $a + b + c$
- Number of non-English speakers: $c + d$
- Number of non-German speakers: $a + d$
- Number of speakers who do not know either language : d
- Number of speakers who know only one of the two languages: $a + c$
- Number of speakers who know at least one of the two languages: $a + b + c$

Example 13.29: In a class of 40 students, 18 students have passed the last math exam, and 24 students have passed the last physics exam. If the number of students who have passed only one of the two exams is 20, then how many students are there who failed the last math exam?

Solution: Let's draw a Venn diagram.

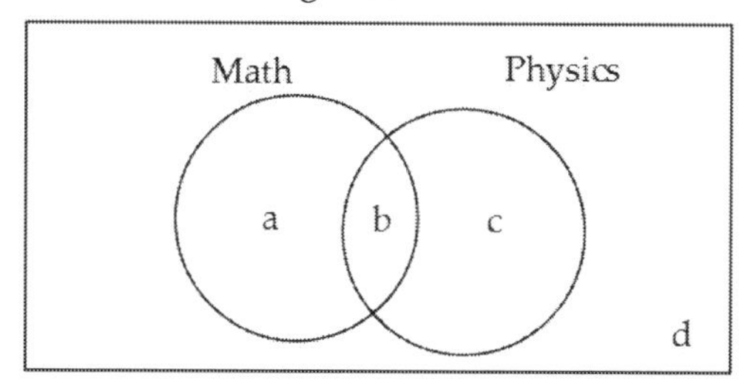

We are given that $a+b+c+d=40$, $a+b=18$, $b+c=24$, and $a+c=20$. Adding the last three equations yields $2a+2b+2c=62$. Dividing this last equation by 2 gives $a+b+c=31$. Substituting 31 for $a+b+c$ in the first equation gives us $31+d=40$, so that we have $d=40-31=9$. We now subtract the equation $a+b=18$ from the equation $a+b+c=31$ to get $c=31-18=13$. So $c+d=13+9=\mathbf{22}$.

Example 13.30: In a class, the number of students who passed the English exam is 24, the number of students who failed the German exam is 16, and the number of students who failed both of the exams is 8. What is the number of students who passed both the English and German exams?

Solution:

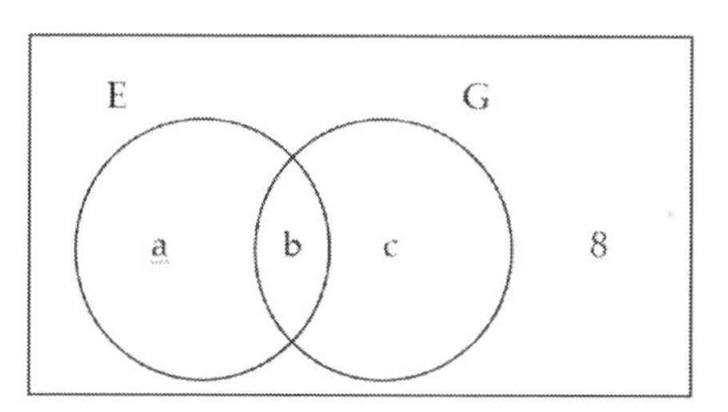

We are given that $a+b=24$, and $a+8=16$. So $a=16-8=8$, and therefore the number of students who passed both the English and German exams is $b=24-a=24-8=\mathbf{16}$.

PROBABILITY

Given an experiment, the **sample space** for the experiment is the set of all possible outcomes. An **event** is a subset of the sample space. The **probability** of an event E, written $P(E)$, is the chance that the event will occur.

Simple Probability Principle: To compute a simple probability where all outcomes are equally likely, divide the number of outcomes in the event E (the "successes") by the total number of outcomes in the sample space.

$$P(E) = \frac{\text{number of outcomes in } E}{\text{total number of outcomes}}$$

Example 13.31: When a 6 sided die is rolled, find the probability that an even number is obtained.

Solution: The sample space for the experiment of rolling a die is $S = \{1,2,3,4,5,6\}$. The event E that an even number is obtained is $E = \{2,4,6\}$. Since all outcomes in the sample space are equally likely, we can use the simple probability principle to get $P(E) = \frac{3}{6} = \frac{1}{2}$.

Example 13.32: When a 6 sided die is rolled, find the probability that a prime number is obtained.

Solution: The event A that a prime number is obtained is $A = \{2,3,5\}$. So $P(A) = \frac{3}{6} = \frac{1}{2}$.

Example 13.33: A 6 sided die is rolled. What is the probability of obtaining a number greater than 4?

Solution: The event B that a number greater than 4 is obtained is $A = \{5,6\}$. So $P(A) = \frac{2}{6} = \frac{1}{3}$.

PROBABILITY AXIOMS

1. For each event A, $0 \leq P(A) \leq 1$.
2. $P(S) = 1$ where S is the sample space of the experiment.
3. **Addition Principle:** $P(A \cup B) = P(A) + P(B) - P(A \cap B)$.
4. A and B are **disjoint** (**mutually exclusive**) events if $A \cap B = \emptyset$ (the empty set). In this case $P(A \cup B) = P(A) + P(B)$.
5. Let A be a subset of the sample space S. The complement of A is $A' = S \setminus A$, and $P(A') = 1 - P(A)$.

CONDITIONAL PROBABILITY

A **conditional probability** is the probability of one event given that another event has already occurred. If we name the events A and B, respectively, then this conditional probability will be written $P(A|B)$. The line "|" is read "given," so that $P(A|B)$ is the probability of A *given* B.

Conditional Probability Formula: $P(A|B) = \frac{P(A \cap B)}{P(B)}$

If $P(A|B) = P(A)$, then we say that A and B are **independent events.**

In this case we have $\boldsymbol{P(A \cap B) = P(A) \cdot P(B)}$ and $P(B|A) = P(B)$.

In applications it is usually easiest to use the definition of independent events given in bold above.

Example 13.34: Let A and B be events with $P(A) = \frac{1}{3}$, $P(B) = \frac{1}{4}$, and $P(A \cap B) = \frac{1}{5}$. Find $P(A|B)$, $P(B|A)$, and $P(A \cup B)$.

Solution: $P(A|B) = \frac{P(A \cap B)}{P(B)} = \frac{1}{5} \div \frac{1}{4} = \frac{1}{5} \cdot \frac{4}{1} = \boldsymbol{\frac{4}{5}}$.

$P(B|A) = \frac{P(B \cap A)}{P(A)} = \frac{1}{5} \div \frac{1}{3} = \frac{1}{5} \cdot \frac{3}{1} = \boldsymbol{\frac{3}{5}}$.

$P(A \cup B) = P(A) + P(B) - P(A \cap B) = \frac{1}{3} + \frac{1}{4} - \frac{1}{5} = \frac{20+15-12}{3 \cdot 4 \cdot 5} = \boldsymbol{\frac{23}{60}}$.

Example 13.35: Let A and B be events with $P(A) = \frac{1}{3}$ and $P(A \cup B) = \frac{1}{2}$. If A and B are mutually exclusive, find $P(B)$.

Solution: Since A and B are mutually exclusive, $A \cap B = \emptyset$, and therefore $P(A \cap B) = 0$. So

$$P(A \cup B) = P(A) + P(B) \Rightarrow \frac{1}{2} = \frac{1}{3} + P(B) \Rightarrow P(B) = \frac{1}{2} - \frac{1}{3} = \boldsymbol{\frac{1}{6}}.$$

Example 13.36: Let A and B be events with $P(A) = \frac{1}{3}$ and $P(A \cup B) = \frac{1}{2}$. If A and B are independent, find $P(B)$.

Solution: Since A and B are independent,

$$P(A \cap B) = P(A) \cdot P(B) = \frac{1}{3} \cdot P(B)$$

So, $P(A \cup B) = P(A) + P(B) - P(A \cap B) \Rightarrow \frac{1}{2} = \frac{1}{3} + P(B) - \frac{1}{3}P(B)$

$$\Rightarrow P(B) - \frac{1}{3}P(B) = \frac{1}{2} - \frac{1}{3} = \frac{1}{6} \Rightarrow \frac{2}{3}P(B) = \frac{1}{6} \Rightarrow P(B) = \frac{1}{6} \cdot \frac{3}{2} = \boldsymbol{\frac{1}{4}}$$

Example 13.37: There are 4 orange, 5 red and 6 blue balls in a bag. Assume that once a ball is drawn it is not placed back into the bag. Two balls are drawn from the bag, one at a time. What is the probability of drawing the first ball orange, and the second ball blue?

Solution: Let A be the event of drawing the first ball orange, and let B be the event of drawing the second ball blue. $P(A) \cdot P(B) = \frac{4}{15} \cdot \frac{6}{14} = \boldsymbol{\frac{4}{35}}$.

Notes: (1) There are a total of $4+5+6=15$ balls in the bag, and 4 of them are orange. Since all probabilities are equally likely, $P(A)=\frac{4}{15}$.

(2) Once an orange ball is selected, there will be a total of 14 balls in the bag, and 6 of them are blue. Since all probabilities are equally likely, $P(B)=\frac{6}{14}$.

(3) The probability of the event A, followed by the event B is $P(A)\cdot P(B)$.

Example 13.38: There are 5 orange, 6 red and 7 blue balls in a bag. Assume that once a ball is drawn it is not placed back into the bag. Two balls are drawn from the bag, one at a time. What is the probability of drawing two blue balls?

Solution: Let A be the event of drawing the first ball blue, and let B be the event of drawing the second ball blue. $P(A)\cdot P(B)=\frac{7}{18}\cdot\frac{6}{17}=\mathbf{\frac{7}{51}}$.

Example 13.39: There are 8 balls in a bag, numbered from 1 to 8. Assume that once a ball is drawn it is not placed back into the bag. Three balls are drawn from the bag, one at a time. What is the probability that all three balls drawn have even numbers?

Solution: Let A, B, and C be the events of getting an even numbered ball on the first, second, and third draws, respectively. Then we have

$$P(A)\cdot P(B)\cdot P(C)=\frac{4}{8}\cdot\frac{3}{7}\cdot\frac{2}{6}=\mathbf{\frac{1}{14}}$$

Example 13.40: A class president and an assistant will be selected among 13 students, 8 boys and 5 girls. Names of the candidates are written on pieces of paper and placed inside a bag. One piece of paper is drawn to determine the president, and then, without replacing the first piece of paper, a second piece of paper is drawn to determine the assistant. Find the probability of selecting a boy class president and a girl assistant.

Solution: Let A be the event of selecting a boy president, and let B be the event of selecting a girl assistant. Then we have

$$P(A)\cdot P(B)=\frac{8}{13}\cdot\frac{5}{12}=\mathbf{\frac{10}{39}}$$

Example 13.41: Two people are randomly selected from 8 men and 6 women. Find the probability that both people chosen are women.

Solution: The total number of ways to choose 2 people from 14 is $C(14,2) = \frac{14!}{12!2!} = \frac{14\cdot13}{2\cdot1} = 91$, and the number of ways to choose 2 women from 6 is $C(6,2) = \frac{6!}{4!2!} = \frac{6\cdot5}{2\cdot1} = 15$.

So the answer is $\mathbf{\frac{15}{91}}$.

Example 13.42: Let A be an event with $P(A) = \frac{3}{5}$. Then $P(A') = ?$

Solution: $P(A') = 1 - P(A) = 1 - \frac{3}{5} = \frac{2}{5}$

Example 13.43: Let A and B be two different events with $P(A) = \frac{1}{2}$, $P(B) = \frac{1}{2}$, and $P(A \cup B) = \frac{3}{4}$. Evaluate $P(A \cap B)$.

Solution: By the addition principle, we have

$$P(A \cup B) = P(A) + P(B) - P(A \cap B)$$

$\frac{3}{4} = \frac{1}{2} + \frac{1}{2} - P(A \cap B) = 1 - P(A \cap B)$. So $P(A \cap B) = 1 - \frac{3}{4} = \mathbf{\frac{1}{4}}$.

Example 13.44: There are 12 marbles in a box. 2 of the marbles are yellow, 5 of the marbles are red, and 5 of the marbles are blue. What is the probability of randomly selecting a red or blue marble from the box?

Solution: Let A be the event of selecting a red marble from the box, and let B be the event of selecting a blue marble from the box. Since $A \cap B = \emptyset$ we have $P(A \cup B) = P(A) + P(B) = \frac{5}{12} + \frac{5}{12} = \frac{10}{12} = \mathbf{\frac{5}{6}}$.

Example 13.45: There are 12 marbles in a box, 8 red and 4 blue. Two marbles are drawn from the box, one at a time, and without replacement. What is the probability that the first marble is blue and second one red?

Solution: Let A be the event that the first marble is blue, and let B be the event that the second marble is red. $P(A) \cdot P(B) = \frac{4}{12} \cdot \frac{8}{11} = \mathbf{\frac{8}{33}}$.

14. FUNCTIONS

A function from the set A to the set B is a rule which assigns each element of A to *exactly* one element of B.

We can use the **vertical line test** to determine if a graph is the graph of a function. If every vertical line intersects the graph in at most one point, the graph is the graph of a function.

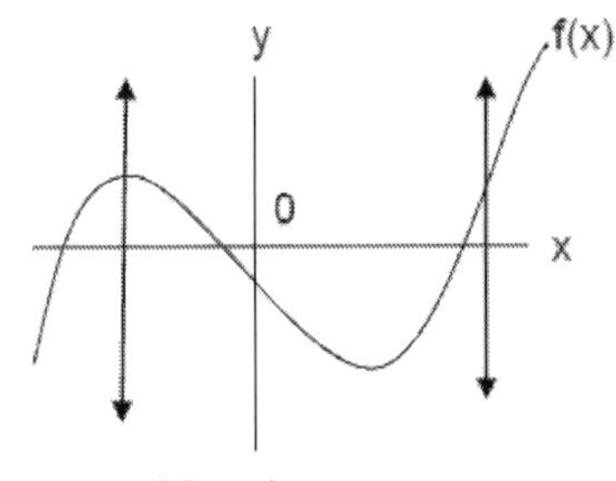

f function

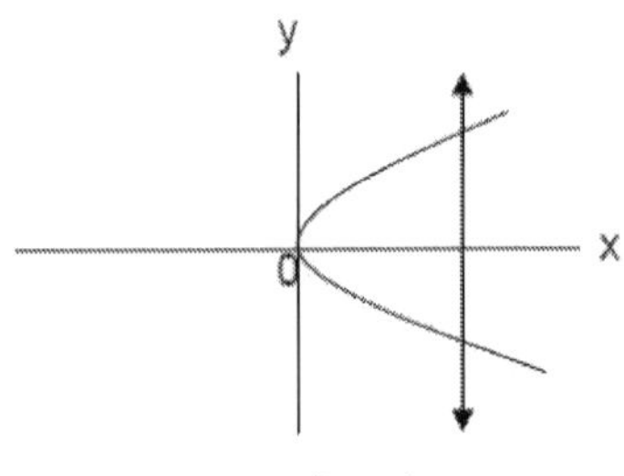

not a function

Example 14.1: If $f(x) = x^2 + 3x + 4$, then $f(1) + f(0) =$

Solution: $f(1) = 1^2 + 3 \cdot 1 + 4 = 1 + 3 + 4 = 8$.

$f(0) = 0^2 + 3 \cdot 0 + 4 = 4$.

So $f(1) + f(0) = 8 + 4 = \mathbf{12}$.

Example 14.2: If $f(x) = 3x + 6$, then $f(2) - f(1) =$

Solution: $f(2) = 3 \cdot 2 + 6 = 6 + 6 = 12$.

$f(1) = 3 \cdot 1 + 6 = 3 + 6 = 9$.

So $f(2) - f(1) = 12 - 9 = \mathbf{3}$.

Example 14.3: If $f(x) = 6x + 7$, then $f(3): f(2) =$

Solution: $f(3) = 6 \cdot 3 + 7 = 18 + 7 = 25$.

$f(2) = 6 \cdot 2 + 7 = 12 + 7 = 19$.

So $f(3): f(2) = \mathbf{\frac{25}{19}}$.

Example 14.4: If $f(x) = \frac{2x+3}{4x+5}$, then $f(3) =$

Solution: $f(3) = \frac{2 \cdot 3+3}{4 \cdot 3+5} = \frac{6+3}{12+5} = \mathbf{\frac{9}{17}}$.

Example 14.5: If $f(x) = 2x + 4$, then $f(x + 1) =$

Solution: $f(x + 1) = 2(x + 1) + 4 = 2x + 2 + 4 = \mathbf{2x + 6}$.

Example 14.6: If $f(x) = 4x + 5$, then $f(x^2 + 1) =$

Solution: $f(x^2 + 1) = 4(x^2 + 1) + 5 = 4x^2 + 4 + 5 = \mathbf{4x^2 + 9}$.

Example 14.7: If $f(x + 1) = (x + 2)f(x)$, and $f(1) = 4$, then $f(4) =$

Solution: When $x = 1$, $f(1 + 1) = (1 + 2)f(1)$, so that $f(2) = 3 \cdot 4 = 12$.

When $x = 2$, $f(2 + 1) = (2 + 2)f(2)$, so that $f(3) = 4 \cdot 12 = 48$.

When $x = 3$, $f(3 + 1) = (3 + 2)f(3)$, so that $f(4) = 5 \cdot 48 = \mathbf{240}$.

LINEAR FUNCTIONS

A function of the form $f(x) = mx + b$ where m and b are real numbers is called a **linear function**.

Example 14.8: If $f(x)$ is a linear function, $f(4) = 9$, and $f(3) = 7$, then $f(5) =$

Quick solution: Since the function is linear, "**equal jumps in x lead to equal jumps in $f(x)$.**" Since the x-jump from 3 to 4 gives an $f(x)$-jump from 7 to 9, an x-jump from 4 to 5 gives an $f(x)$-jump from 9 to 11. So $f(5) = \mathbf{11}$.

Note: By a **jump** in x we mean the difference between two x-values. For example the jump in x when x goes from 3 to 4 is $4 - 3 = 1$. I abbreviated this as an "x-jump." For example, the x-jump from $x = 3$ to $x = 4$ is 1.

Solution using the slope formula: Let's compute the slope of the line. We are given that the points (4,9) and (3,7) are on the line. Therefore the slope of the line is $m = \frac{9-7}{4-3} = \frac{2}{1} = 2$. We can also use the points $(5, f(5))$ and (4,9) to compute the slope. So the slope of the line is

$$2 = \frac{f(5)-9}{5-4} = f(5) - 9. \text{ So } f(5) = 2 + 9 = \mathbf{11}.$$

Direct Solution: Since the function is linear it has the form $f(x) = ax + b$.

Since $f(4) = 9$, $9 = 4a + b$.

Since $f(3) = 7$, $7 = 3a + b$.

Subtracting the second equation from the first gives $2 = a$.

Substituting $a = 2$ into the first equation gives

$$9 = 4 \cdot 2 + b \Rightarrow b = 9 - 8 = 1.$$

So $f(x) = 2x + 1$, and so $f(5) = 2 \cdot 5 + 1 = 10 + 1 = \mathbf{11}$.

The **slope** of the line passing through the points $A(x_1, y_1)$ and $B(x_2, y_2)$ is

$$m = \frac{y_2 - y_1}{x_2 - x_1}$$

Example 14.9: Find the slope of the line passing through the points $A(2,4)$ and $B(-4,8)$.

Solution: $m = \frac{y_2-y_1}{x_2-x_1} = \frac{8-4}{-4-2} = \frac{4}{-6} = -\frac{2}{3}$.

The **point-slope** form of the equation of a line is

$$y - y_1 = m(x - x_1)$$

Where m is the slope of the line and (x_1, y_1) is any point on the line.

If the line passes through the origin (0,0), then the point-slope form gives $y = mx$.

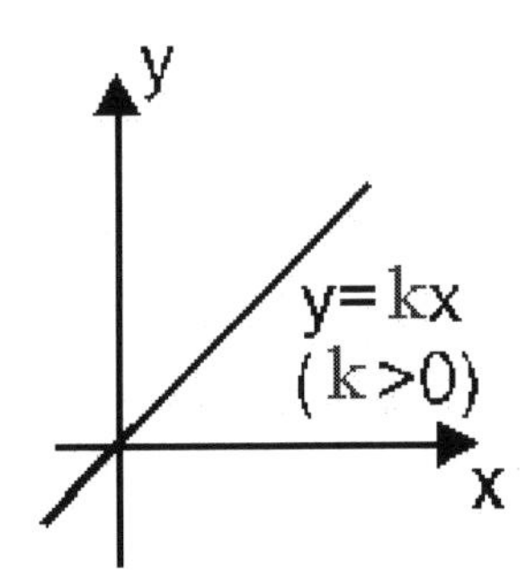

In the figure above, we have a line passing through the origin with slope $m = k > 0$.

If $k < 0$ the line would be moving downward from left to right.

Example 14.10: Sketch the graph of the line with equation $y = 2x$.

Solution:

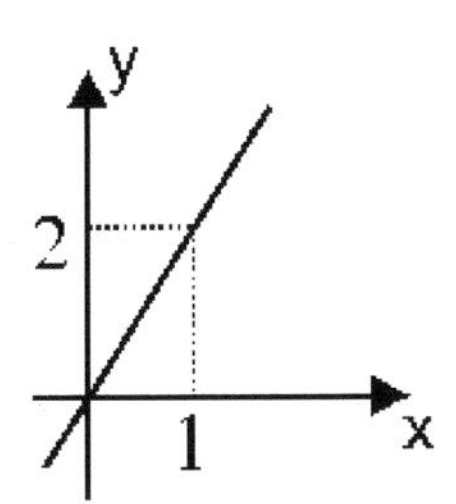

The line passes through (0,0) and the slope of the line is $m = 2 = \frac{2}{1}$.

So we start at (0,0) and then move up 2 and right 1 to plot a second point. Once we have two points we can draw the line.

Example 14.11: Sketch the graph of the line with equation $y = x$.

Solution: $m = 1 = \frac{1}{1}$ (up 1, right 1).

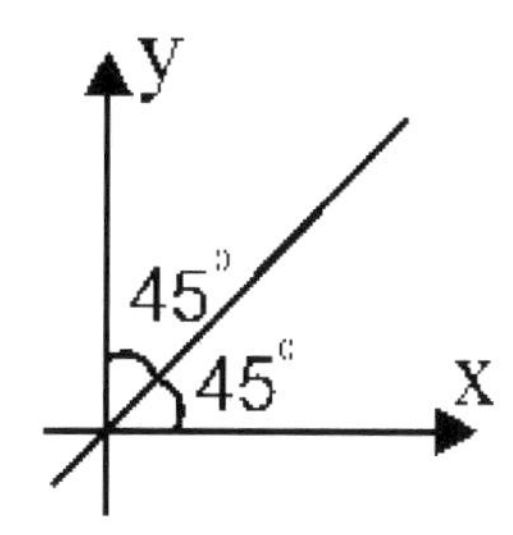

Example 14.12: Sketch the graph of the line with equation $y = -x$.

Solution: $m = -1 = \frac{-1}{1}$ (down 1, right 1).

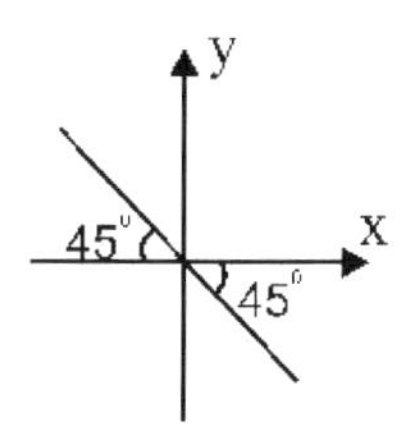

Example 14.13: Write an equation of the line with a slope of 2 that passes through the point $(2, -3)$.

Solution: An equation in point-slope form is $y - (-3) = 2(x - 2)$, or equivalently $\boldsymbol{y + 3 = 2(x - 2)}$.

Note: If we solve the equation for y we get the **slope-intercept** form for the equation of a line: $y = mx + b$.

Here m is the slope, and the point $(0, b)$ is the y-intercept of the line.

In the last example we can distribute the 2 on the right to get $y + 3 = 2x - 4$. Adding 3 to each side of the equation we get $\boldsymbol{y = 2x - 7}$.

Example 14.14: Write an equation of the line passing through the points $(2,3)$ and $(-2,1)$.

Solution: $m = \frac{1-3}{-2-2} = \frac{-2}{-4} = \frac{1}{2}$. So, using the point $(2,3)$, an equation of the line is $\boldsymbol{y - 3 = \frac{1}{2}(x - 2)}$.

Notes: (1) If we use the point $(-2,1)$ instead, we get the equation $\boldsymbol{y - 1 = \frac{1}{2}(x + 2)}$.

(2) Solving either of the equations in point-slope form for y gives the same slope-intercept form: $\boldsymbol{y = \frac{1}{2}x + 2}$.

(3) As another alternative, we have $\frac{y-3}{1-3} = \frac{x-2}{-2-2}$, or equivalently $\frac{y-3}{-2} = \frac{x-2}{-4}$. Cross multiplying gives $-2x + 4 = -4y + 12$, or equivalently $\mathbf{-2x + 4y = 8}$.

(4) The equation of a line in **general form** is $ax + by = c$ where a, b, and c are real numbers with $a \neq 0$ or $b \neq 0$.

The equation in note (3) is written in general form.

The **two-intercept** form of the equation of a line is

$$\frac{x}{a} + \frac{y}{b} = 1$$

where $(a, 0)$ is the x-intercept and $(0, b)$ is the y-intercept.

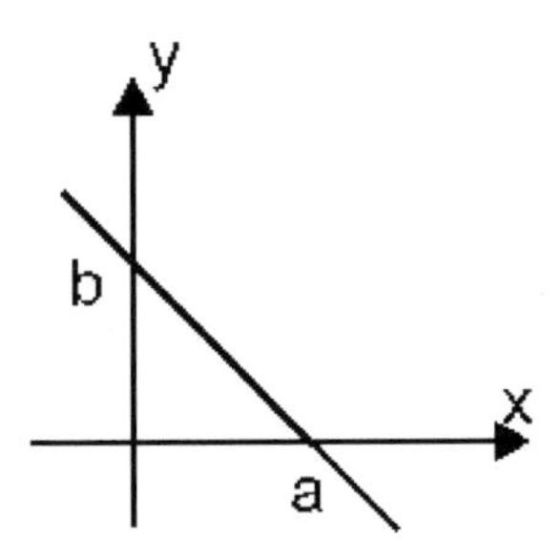

Example 14.15: Write an equation of the line passing through the points (0,4) and (3,0).

Solution:

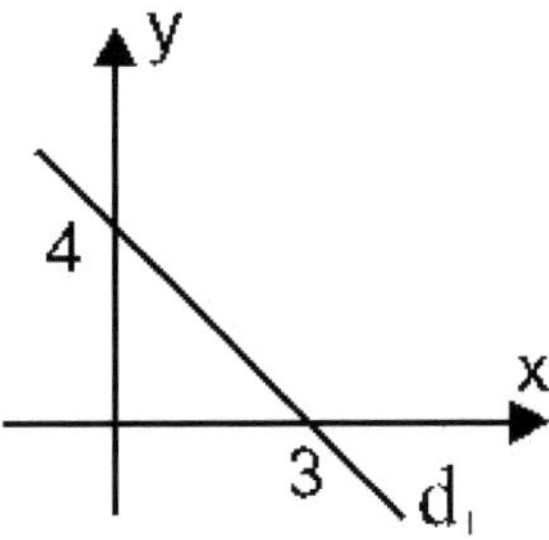

Using the two-intercept form of the equation of the line we get $\frac{x}{3} + \frac{y}{4} = \mathbf{1}$, or equivalently $\mathbf{4x + 3y = 12}$.

Example 14.16: Write an equation of the line passing through the points (0,5) and (−4,0).

Solution:

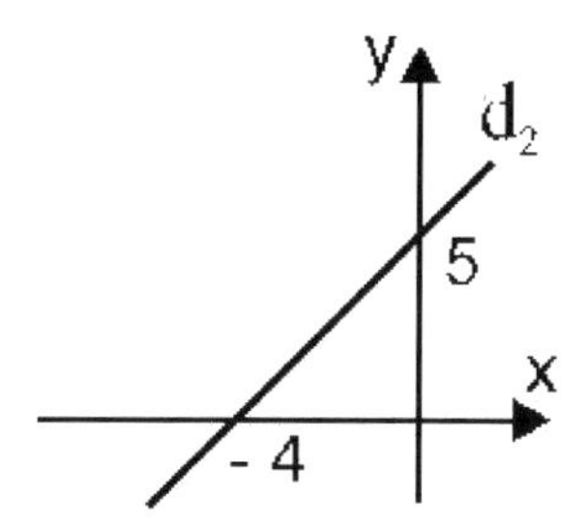

Using the two-intercept form of the equation of the line we get $\frac{x}{-4}+\frac{y}{5}=\mathbf{1}$, or equivalently $-\mathbf{5x}+\mathbf{4y}=\mathbf{20}$.

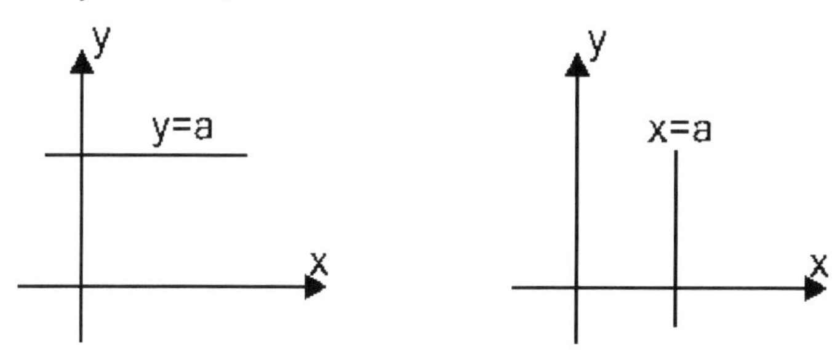

A **horizontal line** has equation $y=a$ for some real number a, and a **vertical line** has equation $x=a$ for some real a.

Example 14.17: Write an equation of the line passing through the points (3,2) and (3,5).

Solution: Both points have 3 as their x-coordinate. The line is therefore a vertical line with equation $\boldsymbol{x}=\mathbf{3}$.

Note: If we were to try to compute the slope of the line we would get $m=\frac{5-2}{3-3}=\frac{3}{0}$ which is undefined. Vertical lines have an **undefined slope**.

Parallel lines have the same slope and perpendicular lines have slopes that are negative reciprocals of each other.

Example 14.18: Write an equation of the line passing through the point (1,5) and parallel to the line with equation $2y+3x=1$

Solution: We solve the given equation for y. We first subtract $3x$ from each side of the equation to get $2y=-3x+1$. We then divide by 2 to get $y=-\frac{3}{2}x+\frac{1}{2}$.

It is now easy to see that the slope of the given line is $m = -\frac{3}{2}$. Since parallel lines have the same slope, we can now write an equation in point-slope form: $\mathbf{y - 5 = -\frac{3}{2}(x - 1)}$.

Example 14.19: Write an equation of the line passing through the point (0,3) and perpendicular to the line with equation $4y - 5x = 8$

Solution: We solve the given equation for y. We first add $5x$ to each side of the equation to get $4y = 5x + 8$. We then divide by 4 to get $y = \frac{5}{4}x + 2$.

It is now easy to see that the slope of the given line is $m = \frac{5}{4}$. Since perpendicular lines have negative reciprocal slopes, we can now write an equation in slope-intercept form: $\mathbf{y = -\frac{4}{5}x + 3}$.

The **general form of an equation of a line** is $\boldsymbol{ax + by = c}$ where a, b and c are real numbers. If $b \neq 0$, then the slope of this line is $m = -\frac{a}{b}$. If $b = 0$, then the line is vertical and has no slope.

Let us consider 2 such equations.

$$\boldsymbol{ax + by = c}$$
$$\boldsymbol{dx + ey = f}$$

(1) If there is a number r such that $ra = d$, $rb = e$, and $rc = f$, then the two equations represent the **same line**. Equivalently, the two equations represent the same line if $\frac{a}{d} = \frac{b}{e} = \frac{c}{f}$. In this case the system of equations has **infinitely many solutions**.

(2) If there is a number r such that $ra = d$, $rb = e$, but $rc \neq f$, then the two equations represent **parallel** but distinct lines. Equivalently, the two equations represent parallel but distinct lines if $\frac{a}{d} = \frac{b}{e} \neq \frac{c}{f}$. In this case the system of equations has **no solution**.

(3) Otherwise the two lines intersect in a single point. In this case $\frac{a}{d} \neq \frac{b}{e}$, and the system of equations has a **unique solution**.

These three cases are illustrated in the figure below.

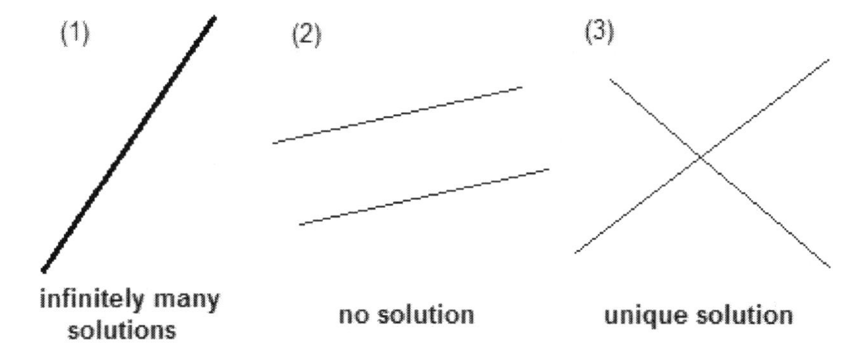

Example 14.20: The following two equations represent the same line.

$$2x + 8y = 6$$
$$3x + 12y = 9$$

To see this note that $\frac{2}{3} = \frac{8}{12} = \frac{6}{9}$ (or equivalently, let $r = \frac{3}{2}$ and note that $\left(\frac{3}{2}\right)(2) = 3$, $\left(\frac{3}{2}\right)(8) = 12$, and $\left(\frac{3}{2}\right)(6) = 9$).

The following two equations represent parallel but distinct lines.

$$2x + 8y = 6$$
$$3x + 12y = 10$$

This time $\frac{2}{3} = \frac{8}{12} \neq \frac{6}{10}$.

The following two equations represent a pair of intersecting lines.

$$2x + 8y = 6$$
$$3x + 10y = 9$$

This time $\frac{2}{3} \neq \frac{8}{10}$.

RECTANGULAR COORDINATES

We form the **rectangular** (or **Cartesian**) **coordinate system** by intersecting a vertical line with a horizontal line. The horizontal line is called the ***x*-axis**, and the vertical line is called the ***y*-axis**. The point of intersection of these two lines is called the **origin**, and it has coordinates (0,0).

A point has the form (x, y). The real number x is called the **x-coordinate** of the point and the real number y is called the **y-coordinate** of the point. To plot the point (x, y) we move right or left $|x|$units from the origin and then up or down $|y|$ units from the origin, depending upon if x and y are positive or negative.

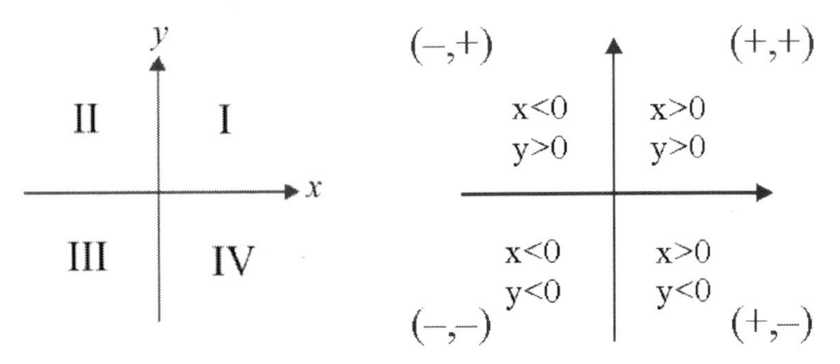

The x- and y-axes split the **xy-plane** into four quadrants. These quadrants are labelled in the figure above on the left. The figure on the right shows which coordinates are positive and negative in each quadrant. For example, in quadrant I, both the x- and y-coordinates of each point are positive, whereas in quadrant II, the x-coordinate of each point is negative, and the y-coordinate of each point is positive.

The **distance** between the points $A(x_1, y_1)$ and $B(x_2, y_2)$ in the plane is

$$d(A, B) = \sqrt{(x_2 - x_1)^2 + (y_2 - y_1)^2}$$

The following figure shows a geometric justification for this formula.

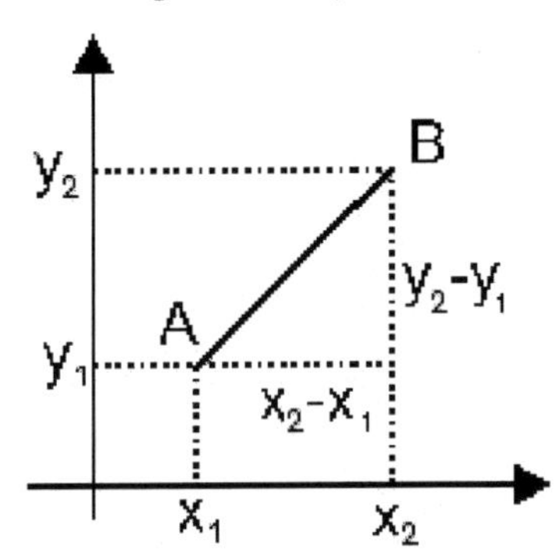

Note that the distance between A and B is simply the length of the line segment AB. By the Pythagorean Theorem, we see that

$$AB^2 = (x_2 - x_1)^2 + (y_2 - y_1)^2.$$

The distance formula follows from this observation.

The distance between the points $A(x_1, y_1, z_1)$ and $B(x_2, y_2, z_2)$ in space is

$$d(A,B) = \sqrt{(x_2 - x_1)^2 + (y_2 - y_1)^2 + (z_2 - z_1)^2}$$

Example 14.21: Find the distance between the points $A(-1,7)$ and $B(3,4)$.

Solution: $d(A,B) = \sqrt{\left(3 - (-1)\right)^2 + (4 - 7)^2} = \sqrt{4^2 + (-3)^2}$

$$= \sqrt{16 + 9} = \sqrt{25} = \mathbf{5}.$$

The **midpoint** of the line segment with endpoints $A(x_1, y_1)$ and $B(x_2, y_2)$ is

$$M = \left(\frac{x_1 + x_2}{2}, \frac{y_1 + y_2}{2}\right)$$

Note that to get the x-coordinate of the midpoint we are simply taking the arithmetic mean of the x-coordinates of the two points. Similarly, to get the y-coordinate of the midpoint we are taking the arithmetic mean of the y-coordinates of the two points.

Area and Centroid of a Triangle

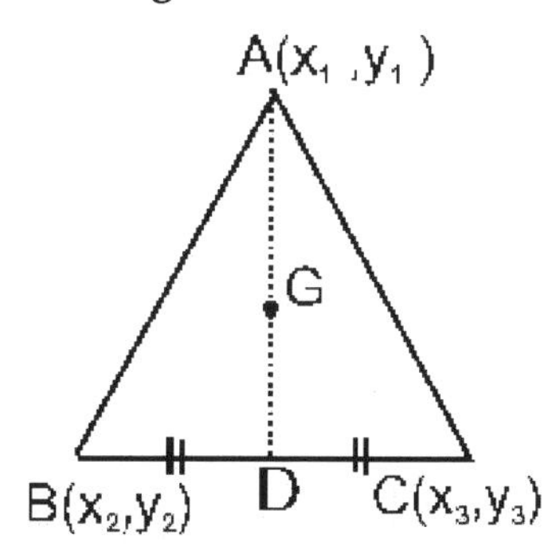

In the figure above, $A(x_1, y_1)$, $B(x_2, y_2)$, and $C(x_3, y_3)$ are vertices of ΔABC, G is the **centroid** (or **center of gravity**) of the triangle, and D is the midpoint of side BC. We have

$$D = \left(\frac{x_2 + x_3}{2}, \frac{y_2 + y_3}{2}\right) \qquad G = \left(\frac{x_1 + x_2 + x_3}{3}, \frac{y_1 + y_2 + y_3}{3}\right)$$

$$\text{Area of } \Delta ABC = \frac{1}{2}[(x_1y_2 + x_2y_3 + x_3y_1) - (x_1y_3 + x_3y_2 + x_2y_1)]$$

The following figure may help to visualize this area formula:

$$\begin{vmatrix} x_1 & y_1 \\ x_2 & y_2 \\ x_3 & y_3 \\ x_1 & y_1 \end{vmatrix}$$

Example 14.22: Find the centroid and area of the triangle with vertices $A(7,2)$, $B(1,3)$, and $C(4,1)$.

Solution: The centroid is $\left(\frac{7+1+4}{3}, \frac{2+3+1}{3}\right) = \left(\frac{12}{3}, \frac{6}{3}\right) = \mathbf{(4, 2)}$.

The area is $\frac{1}{2}(21 + 1 + 8 - 7 - 12 - 2) = \frac{9}{2}$.

QUADRATIC FUNCTIONS

The **general form** for a quadratic function is

$$\boldsymbol{y = ax^2 + bx + c},$$

where a, b, and c are real numbers with $a \neq 0$.

The graph of this function is a parabola whose vertex has x-coordinate

$$-\frac{\boldsymbol{b}}{\boldsymbol{2a}}$$

- If $a > 0$, the parabola opens upward. ∪
- If $a < 0$, the parabola opens downward. ∩
- The vertex has coordinates $\left(-\frac{b}{2a}, f(-\frac{b}{2a})\right) = (-\frac{b}{2a}, \frac{4ac-b^2}{4a})$

The **discriminant** of the quadratic function $y = ax^2 + bx + c$ is

$$\Delta = b^2 - 4ac$$

If $\Delta > 0$, the parabola intersects the x-axis at two distinct points.
If $\Delta = 0$, the parabola is tangent to the x-axis..
If $\Delta < 0$, the parabola does not intersect the x-axis.

Example 14.23: Let's draw the graph of $y = ax^2$ where a is a real number.

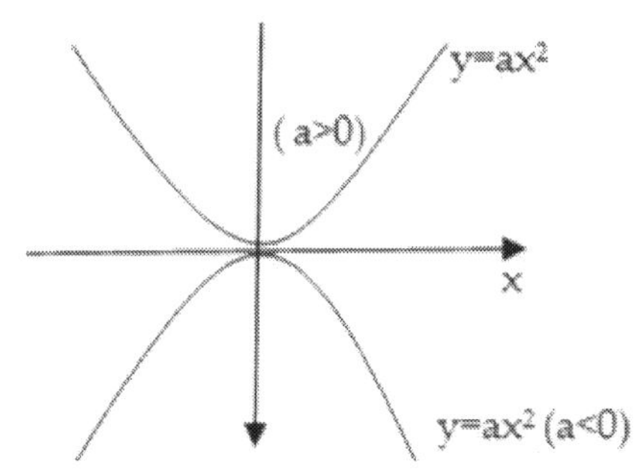

In the figure above we drew two parabolas: one for the case where $a > 0$, and one for the case where $a < 0$.

Note that in both cases, the vertex of the parabola is the point (0,0).

Example 14.24: Let's draw the graph of $y = ax^2 + c$ where a and c are positive real numbers

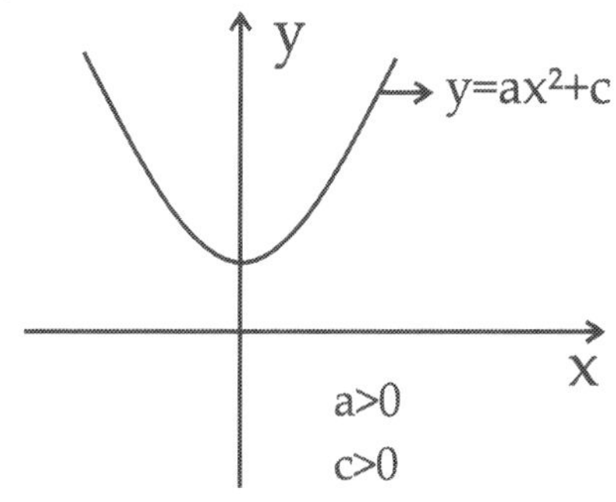

Example 14.25: Draw the graph of $y = x^2 + 2x - 8$.

Solution: We have $a = 1$, $b = 2$, and $c = -8$.

Since $a > 0$, the parabola opens upwards. ($\cup$)

$-\frac{b}{2a} = -\frac{2}{2\cdot 1} = -1$ and when $x = -1$, $y = (-1)^2 + 2(-1) - 8 = -9$. So the vertex of the parabola is $(-1, -9)$.

$y = x^2 + 2x - 8 = (x + 4)(x - 2)$. So the x-intercepts are $(-4,0)$ and $(2,0)$.

When $x = 0$, $y = -8$. So the y-intercept is $(0, -8)$.

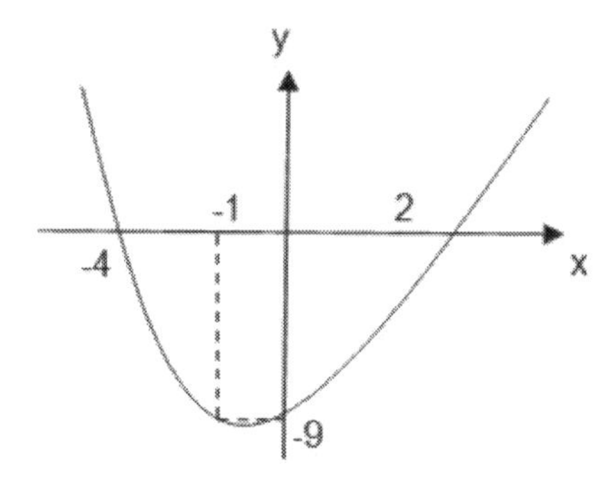

The standard form for a quadratic function is

$$\boldsymbol{y-k=a(x-h)^2} \quad \text{or} \quad \boldsymbol{y=a(x-h)^2+k}$$

The graph is a parabola with **vertex** at (h,k). The parabola opens upwards if $a > 0$ and downwards if $a < 0$.

Example 14.26: Draw the graph of $y = 4(x-3)^2 - 2$.

Solution: Since $a = 4 > 0$, the parabola opens upward. The vertex of the parabola is $(3,-2)$.

$$y = 0 \Leftrightarrow 4(x-3)^2 - 2 = 0 \Leftrightarrow 4(x-3)^2 = 2 \Leftrightarrow (x-3)^2 = \frac{2}{4} = \frac{1}{2}$$

$$\Leftrightarrow x - 3 = \pm\frac{1}{\sqrt{2}} \Leftrightarrow x = 3 \pm \frac{1}{\sqrt{2}}.$$

So the x-intercepts are $(3-\frac{1}{\sqrt{2}},0)$ and $(3+\frac{1}{\sqrt{2}},0)$.

When $x = 0$, $y = 4(-3)^2 - 2 = 4\cdot 9 - 2 = 36 - 2 = 34$. So the y-intercept is $(0,34)$.

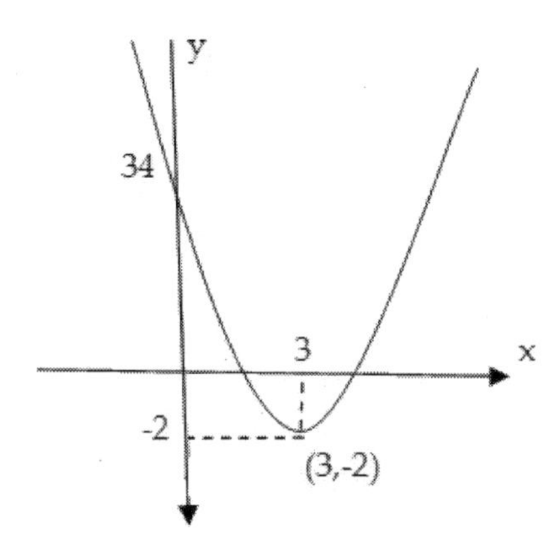

Example 14.27: Draw the graph of $y = -5x^2$

Solution: Since $a = -5 < 0$, the parabola opens downward. The vertex, x- and y-intercept of the parabola is (0,0).

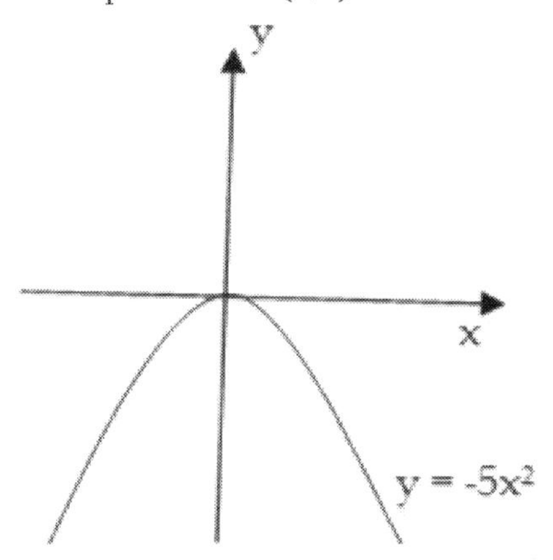

Example 14.28: Draw the graph of $y = x^2 + 3$.

Solution: Since $a = 1 > 0$, the parabola opens upwards. The vertex and, y-intercept of the parabola is (0,3).

$y = 0 \Leftrightarrow x^2 + 3 = 0$. Since this last equation has no real solutions, the parabola has no x-intercepts.

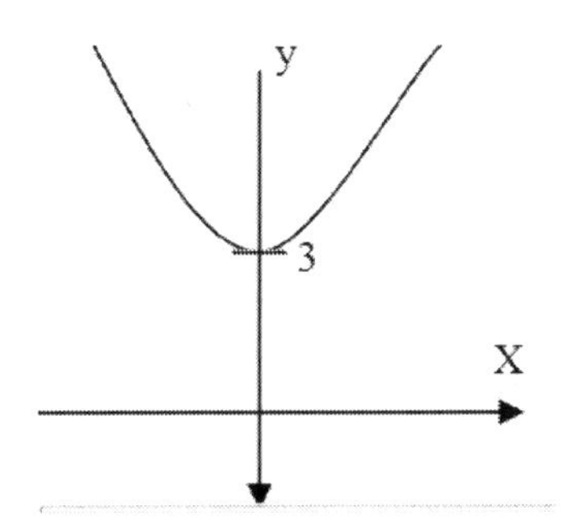

Example 14.29: Draw the graph of $y = 16 - x^2$.

Solution: We have $a = -1$, $b = 0$, and $c = 16$.

Since $a < 0$, the parabola opens downwards.

The vertex and y-intercept of the parabola is (0,16).

$y = 16 - x^2 = (4 + x)(4 - x)$. So the x-intercepts are $(-4,0)$ and $(4,0)$.

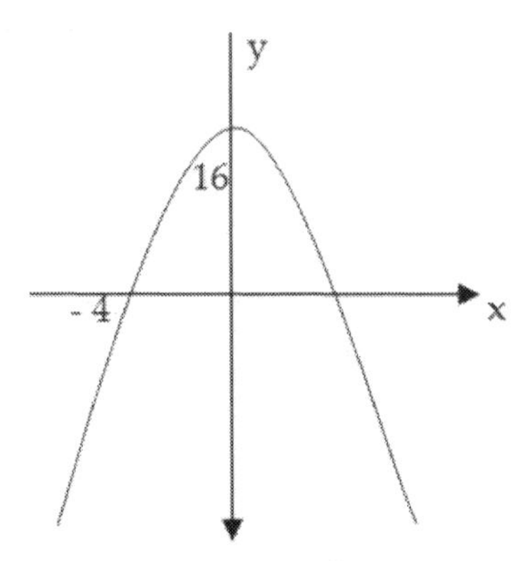

Example 14.30: Draw the graph of $y = x^2 + 6x$.

Solution: We have $a = 1$, $b = 6$, and $c = 0$.

Since $a > 0$, the parabola opens upwards.

$-\frac{b}{2a} = -\frac{6}{2 \cdot 1} = -3$ and when $x = -3$, $y = (-3)^2 + 6(-3) = 9 - 18 = -9$. So the vertex of the parabola is $(-3, -9)$.

$y = x^2 + 6x = x(x + 6)$. So the x-intercepts are $(-6,0)$ and $(0,0)$.

When $x = 0$, $y = 0$. So the y-intercept is $(0,0)$.

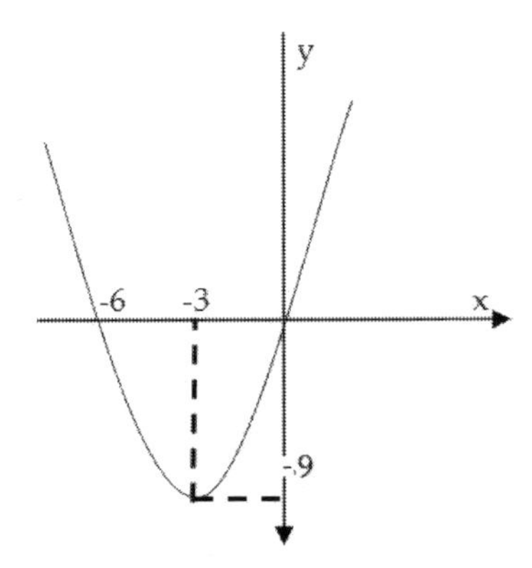

Example 14.31: Draw the graph of $y = (x - 3)^2$.

Solution: Since $a = 1 > 0$, the parabola opens upward. The vertex of the parabola is $(3,0)$.

$$y = 0 \Leftrightarrow (x - 3)^2 = 0 \Leftrightarrow x = 3$$

So there is just one x-intercept of $(3,0)$.

When $x = 0$, $y = (-3)^2 = 9$. So the y-intercept is $(0,9)$.

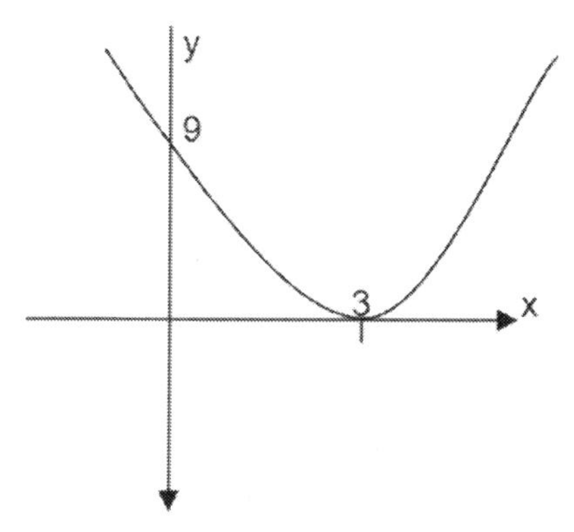

Example 14.32: Draw the graph of $y = -4x^2 + 4$.

Solution: We have $a = -4$, $b = 0$, and $c = 4$.

Since $a = -4 < 0$, the parabola opens downwards.

The vertex and y-intercept of the parabola is $(0,4)$.

$y = -4x^2 + 4 = -4(x^2 - 1) = -4(x + 1)(x - 1)$. So the x-intercepts are $(-1,0)$ and $(1,0)$.

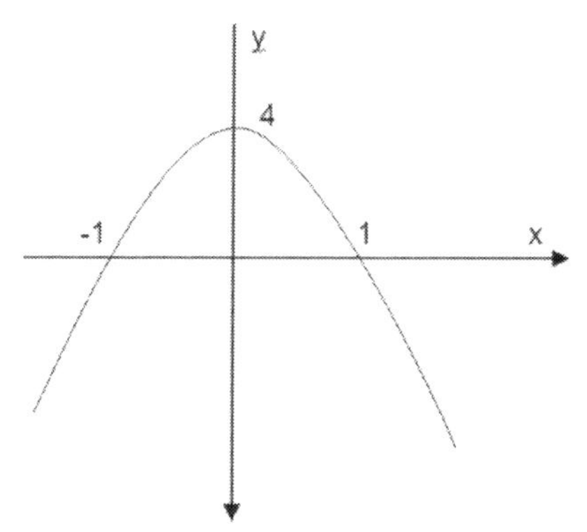

15. MODULAR ARITHMETIC

Let x, y, and n be integers with $n > 1$. We write $x \equiv_n y$ or $x \equiv y \pmod{n}$ if x and y have the same remainder when divided by n.

In this case we say that ***x* and *y* are congruent modulo *n***.

An equivalent way to say that x and y are congruent modulo n is that $x - y$ is divisible by n. In other words, there is an integer k such that $x - y = nk$.

Example 15.1: $4 \equiv_2 10$ because $10 - 4 = 6$, and 6 is divisible by 2.

Equivalently, both 4 and 10 give a remainder of 0 when divided by 2.

Example 15.2: $3 \equiv_2 15$ because $15 - 3 = 12$, and 12 is divisible by 2.

Equivalently, both 3 and 15 give a remainder of 1 when divided by 2.

Example 15.3: If j and k are even, then $j \equiv_2 k$ because $k - j$ is even.

If j and k are odd, then $j \equiv_2 k$ because $k - j$ is even.

Example 15.4: $8 \equiv_3 26$ because $26 - 8 = 18$, and 18 is divisible by 3.

Equivalently, both 8 and 26 give a remainder of 2 when divided by 3.

- When dividing an integer m by the integer $n > 1$, the remainder must be 0, 1, 2, 3, 4, …, or (n – 1). We let $[m]_n$ be this remainder. If it is clear what n is, then we can use the simpler notation $\overline{m}$.
- $[m]_n = \overline{m}$ is called the **congruence class of *m* modulo *n***.
- Z/n is the set of congruence classes modulo n.

So $Z/n = \{\overline{0}, \overline{1}, \dots, \overline{n-1}\}$.

Example 15.5: $[4]_2 = [10]_2 = \overline{0}$. In fact, if j is any even integer, then $[j]_2 = \overline{0}$.

Similarly, if k is an odd integer, then $[k]_2 = \overline{1}$.

Example 15.6: $Z/2 = \{\overline{0}, \overline{1}\}$

Example 15.7: $Z/3 = \{\overline{0}, \overline{1}, \overline{2}\}$

If $x \equiv_n y$ and $w \equiv_n v$, then we have the following:

1. $x + w \equiv_n y + v$
2. $x - w \equiv_n y - v$
3. $xw \equiv_n yv$
4. $x^k \equiv_n y^k$
5. $x - y \equiv_n 0$
6. $kx \equiv_n ky$

Example 15.8: $10 \equiv_4 2$ because $2 - 10 = -8$, and -8 is divisible by 4.

Equivalently, both 10 and 2 give a remainder of 2 when divided by 4.

Example 15.9: $27 \equiv_6 3$ because $3 - 27 = -24$, and -24 is divisible by 6.

Equivalently, both 27 and 3 give a remainder of 3 when divided by 6.

Example 15.10: If $34 \equiv_5 x$, then what is the least positive integer solution x?

Solution: The remainder when we divide 34 by 5 is 4. So $x =$ **4**.

Example 15.11: If $45 \equiv_7 x$, then what is the least positive integer solution x?

Solution: The remainder when we divide 45 by 7 is 3. So $x =$ **3**.

Example 15.12: If today is Wednesday, then what day of the week will it be in **96** days?

Solution: When we divide 96 by 7 the remainder is 5. So $96 \equiv_7 5$.

The fifth day after Wednesday is **Monday**.

Example 15.13: What is the remainder when we divide 2^{26} by 7 ?

Solution: $2^1 = 2$. So $2^1 \equiv_7 2$.

$2^2 = 4$. So $2^2 \equiv_7 4$.

$2^3 = 8$. So $2^3 \equiv_7 1$.

$2^4 = 16$. So $2^4 \equiv_7 2$.

$2^5 = 32$. So $2^5 \equiv_7 4$.

$2^6 = 64$. So $2^6 \equiv_7 1$.

$26 \equiv_3 2$. So $2^{26} \equiv_7 2^2 \equiv_7$ **4**.

Example 15.14: What is the remainder when we divide 5^{63} by 6 ?

Solution: $5^1 = 5$. So $5^1 \equiv_6 5$.

$5^2 = 25$. So $5^2 \equiv_6 1$.

$5^3 = 125$. So $5^3 \equiv_6 5$.

$63 \equiv_2 1$. So $5^{63} \equiv_6 5^1 \equiv_6$ **5**.

Example 15.15: A soldier stands guard every 5 days. He stands his fourth guard on Thursday. When will he keep his twelfth guard?

Solution: If each guard is five days after the last, then $8 \cdot 5 = 40$ days will pass after his fourth guard until his twelfth guard. Since $40 \equiv_7 5$, he will keep his twelfth guard five days after Thursday:

Friday	Saturday	Sunday	Monday	Tuesday
1	2	3	4	5

So he will keep his twelfth guard on **Tuesday**.

16. TABLES AND GRAPHS

Example 16.1: Let's analyze the following table.

Table 1

Grade	Foreign Language
6th grade	20
7th grade	18
8th grade	24
9th grade	26
10th grade	22
11th grade	24
12th grade	30

The table above shows the number of secondary and high school students that can speak a foreign language.

- The 12th grade has the most students that can speak a foreign language.
- The number of secondary school students that can speak a foreign language is $20 + 18 + 24 = 62$.
- The number of high school students that can speak a foreign language is $26 + 22 + 24 + 30 = 102$.
- According to the table, the percentage of students that speak a foreign language that are secondary school students is

$$\frac{62}{62 + 102} = \frac{62}{164} \approx .378 = 37.8\%$$

- According to the table, the percentage of students that speak a foreign language that are high school students is

$$\frac{102}{62 + 102} = \frac{102}{164} \approx .62195 \approx 62.2\%$$

Example 16.2:

Table 2

Students by month born		
Month	Female	Male
January	12	16
February	24	20
March	18	26
April	22	28

The table above shows the number of students from a certain school who were born in January, February, March, and April.

- The number of the students born in January is $12 + 16 = 28$.
- The number of the students born in February is $24 + 20 = 44$.
- The number of the students born in March is $18 + 26 = 44$.
- The number of the students born in April is $22 + 28 = 50$.
- The number of female students from the school born in January, February, March, or April is $12 + 24 + 18 + 22 = 76$.
- The number of male students from the school born in January, February, March, or April is $16 + 20 + 26 + 28 = 90$.
- The ratio of the female students to male students born in the first four months is $76 : 90$
- The percentage of male students from the school born in the first four months is $\frac{90}{90+76} = \frac{90}{166} = \frac{45}{83} \approx .542 = 54.2\%$.

Example 16.3:

Table 3

Class	Questions	Correct Answers
Physics	50	46
Chemistry	50	40
Biology	50	42
Mathematics	50	46

The table above shows the test results of a student from four classes: Physics, Chemistry, Biology, and Mathematics. Each test consisted of 50 questions.

- The student answered 46 questions correctly on the Physics test, 40 questions correctly on the Chemistry test, 42 questions correctly on the Biology test, and 46 questions correctly on the Mathematics test.
- The student got the most correct in Physics and Mathematics.
- The student got the least correct in Chemistry.
- The student answered 6 more questions correctly in Physics and Mathematics than he did in Chemistry.

Example 16.4

Table 4

Day	Tourists
Monday	600
Tuesday	200
Wednesday	800
Thursday	400
Friday	1000

The table above shows the number of tourists that visit a certain museum over a period of five days. Let's sketch the data as a **line graph**.

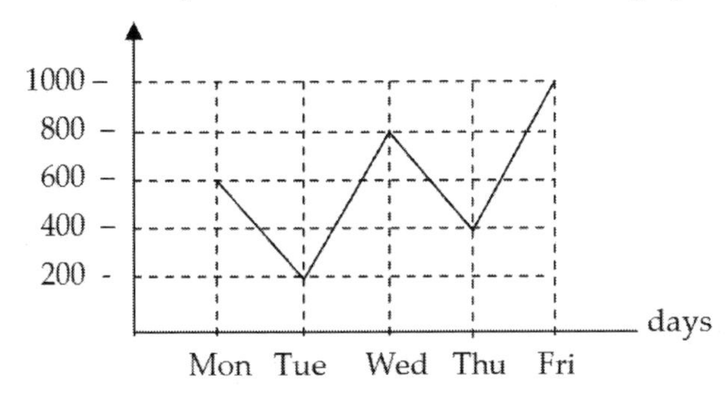

Example 16.5: A bookstore sold 40 books on Monday, 30 on Tuesday, 20 on Wednesday, 25 on Thursday, 40 on Friday and 15 on Saturday. Let's draw the **vertical column chart** (or **vertical bar chart**).

Solution:

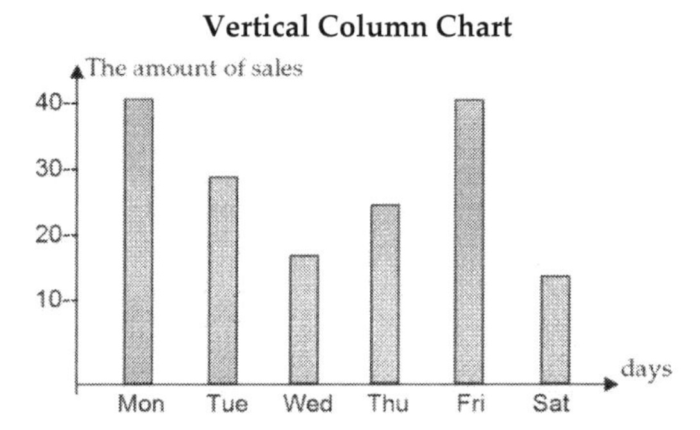

Example 16.5:

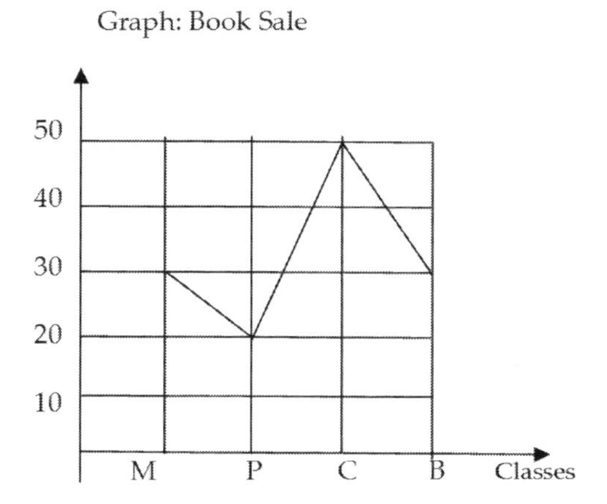

The graph above shows the test results of a student on four exams.

M = Math, P = Physics, C = Chemistry, B = Biology

A) How many answers did the student get correct in each subject?

Solution: He got 30 Math questions correct, 20 Physics questions correct, 50 Chemistry questions correct, and 30 Biology questions correct, for a total of 130 questions.

B) What is the arithmetic mean of the number of correct answers the student received from the four exams?

$$\frac{30+20+50+30}{4}=\frac{130}{4}=\mathbf{\frac{65}{2}}=\mathbf{32.5}$$

Example 16.6:

Table 5

Monday	Tuesday	Wednesday	Thursday	Friday
30 minutes	40 minutes	25 minutes	35 minutes	50 minutes

The above table shows the amount of time John spent reading during each weekday of a specific week. Let's draw the **horizontal column chart** (or **horizontal bar chart)**.

Solution:

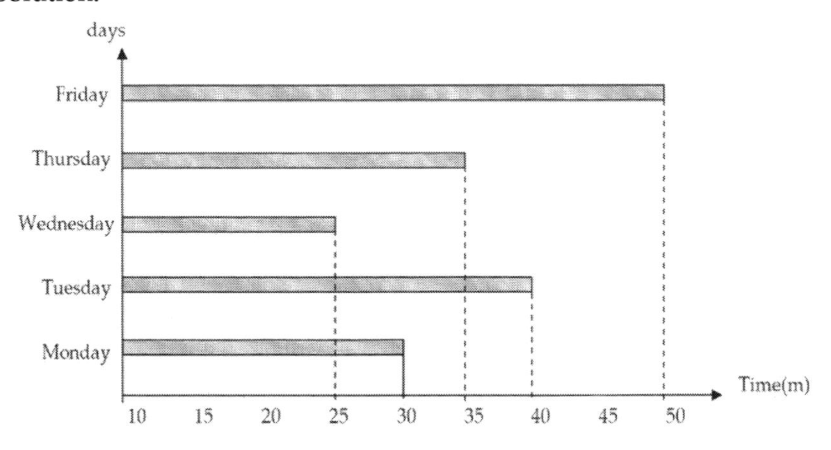

What is John's average reading time for the 5 days?

$$\frac{30+40+25+35+50}{5}=\frac{180}{5}=\mathbf{36}$$

Example 16.7:

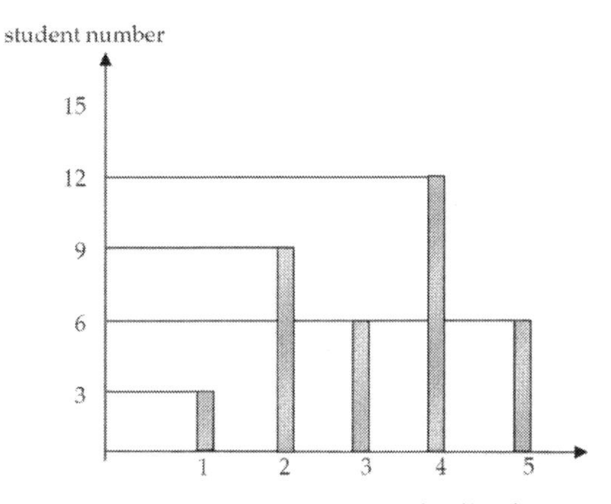

Graph of the exam score distribution

The bar chart above shows the scores of 36 students on a math exam. For example, three students scored a 1 on the exam. According to the chart, what is the arithmetic mean of the 36 scores?

Solution: Arithmetic mean = $\frac{\text{sum of scores}}{\text{total number of students}}$

$$\frac{3 \cdot 1 + 9 \cdot 2 + 6 \cdot 3 + 12 \cdot 4 + 6 \cdot 5}{(3 + 9 + 6 + 12 + 6)} = \frac{117}{36} = \mathbf{\frac{13}{4}} = \mathbf{3.25}$$

17. ALGEBRAIC EQUATIONS

TRANSLATING BETWEEN MATH AND ENGLISH

The expression $x + m$ can be read as:

- x increased by m
- m added to x
- m more than x
- the sum of x and m

The expression $x - m$ can be read as:

- x decreased by m
- m subtracted from x
- m less than x
- the difference between x and m

(I've never liked the last expression because the language doesn't really make it clear if x or m should be written first. But it is standard to write "the difference between x and m" as $x - m$.)

The expression $x \cdot m$ can be read as:

- x times m
- m multiplied by x
- the product of x and m

$\frac{a}{b} \cdot x$ can be read as "$\frac{a}{b}$ of x."

$n(x + m)$ can be read as "n times the sum of m and x."

$\frac{a}{b}(x - m)$ can be read as "$\frac{a}{b}$ of the difference between x and m."

$\frac{a}{b} \cdot \frac{c}{d} \cdot x$ can be read as "$\frac{a}{b}$ of $\frac{c}{d}$ of x."

Example 17.1: The product of a number and the difference between $\frac{1}{2}$ and $\frac{1}{3}$ is 10. What is the number?

Solution: Let x be the number. Then $x\left(\frac{1}{2} - \frac{1}{3}\right) = 10$. Multiplying each side of this equation by 6 gives $x(3 - 2) = 60$, or $x = \mathbf{60}$.

Example 17.2: A number times the sum of $\frac{1}{2}$ and $\frac{1}{3}$ is 75. What is the number?

Solution: Let x be the number. Then $x\left(\frac{1}{2} + \frac{1}{3}\right) = 75$. Multiplying each side of this equation by 6 gives $x(3 + 2) = 450$, or $5x = 450$. Therefore $x = \frac{450}{5} = \mathbf{90}$.

Example 17.3: The first half of a bar is divided into 15 equal pieces, and the second half of the bar is divided into 12 equal pieces. If the positive difference between two different size pieces is 4 cm, then find the length of the bar.

Solution:

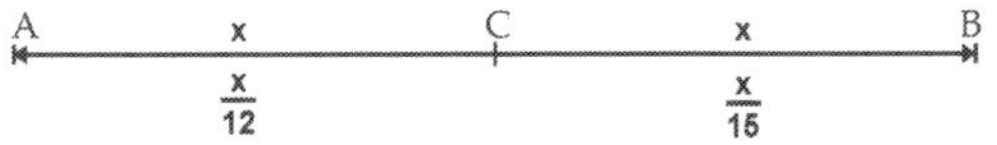

A x C x B

$\frac{x}{12}$ $\frac{x}{15}$

If we let $2x$ be the length of the bar, then we are given that $\frac{x}{12} - \frac{x}{15} = 4$. Multiplying each side of this equation by 60 gives $5x - 4x = 240$, or $x = 240$. So $2x = 2 \cdot 240 = \mathbf{480\ cm}$.

Example 17.4: Ahmet is reading a book with 225 pages. Each day he will read 3 more pages than he did the previous day. If he finishes the book in 10 days, how many pages will he have read on the fourth day?

Solution: 1st day : 2nd day : 3rd day :.............................. 10th day

x x+3 x+6 x+27

$$x + (x+3) + (x+6) + \cdots (x+27) = 225$$

$$10x + 9 \cdot \frac{3+27}{2} = 225$$

$$10x + 135 = 225$$

$$10x = 90$$

$$x = 9$$

On the fourth day, Ahmet will have read $x + 9 = 9 + 9 = \mathbf{18}$ pages.

Note: $3 + 6 + 9 + 12 + 15 + 18 + 21 + 24 + 27$ is an arithmetic series. There are 9 terms, and the average of the first and last term is

$$\frac{3+27}{2} = 135$$

See section 4 for more information on arithmetic series.

Example 17.5: When $\frac{2}{7}$ of a bar is cut off, the midpoint is displaced 14 cm. What was the original length of the bar?

Solution: Let x be the original length of the bar. Then we are given that $\left(x - \frac{2x}{7}\right) \div 2 = \frac{x}{2} - 14$. Multiplying each side of this equation by 14 gives $7x - 2x = 7x - 196$. So $2x = 196$, and $x = \frac{196}{2} = \mathbf{98\ cm.}$

Example 17.6: If $\frac{3}{7}$ of a number is 30, than what is the value of the number?

Solution: Let us call the number x. Then $\frac{3}{7}x = 30$. So $x = 30 \cdot \frac{7}{3} = \mathbf{70}$.

Example 17.7: A parking area can fit 5 trucks and 17 cars. It can also fit 4 trucks and 23 cars. How many cars can park in this parking area?

Solution: Let x be the size of a truck, and let y be the size of a car. We have that

$5x + 17y = 4x + 23y$ or $x = 6y$. Therefore

$$5x + 17y = 5(6y) + 17y = 30y + 17y = 47y.$$

So 47 cars can park in the parking area.

Example 17.8: A customer pays $31 for 3 cans of coke, 4 cans of sprite, and 5 bottles of water. If the price of a can of sprite is 2 times the price of a can of coke and the price of water is 4 times the price of a can of coke, then what is the price of a can of sprite?

Solution:

	coke:	sprite:	water
price:	x	2x	4x

$$3x + 4(2x) + 5(4x) = 31$$
$$3x + 8x + 20x = 31$$
$$31x = 31$$
$$x = \frac{31}{31} = 1$$

So the price of a can of sprite is $2 \cdot 1 =$ **2 dollars**.

Example 17.9: On a certain game show competitors receive 30 points for each correct answer, and lose 40 points for each wrong answer. One of the competitors answers 50 questions, and winds up with 800 points. How many questions did the competitor answer correctly?

Solution: Let x be the number of questions that the competitor answered correctly. It follows that $50 - x$ is the number of questions that the competitor answered wrong. We have $30x - 40(50 - x) = 800$. Distributing the -40 on the left gives $30x - 2000 + 40x = 800$, or equivalently $70x - 2000 = 800$. So $70x = 2800$, and $x = \frac{2800}{70} = \mathbf{40}$.

SETTING UP EQUATIONS INVOLVING FRACTIONS

Example 17.10: Find a of x where a is the sum of $\frac{1}{3}$ and $\frac{2}{7}$.

Solution: $a \cdot x = \left(\frac{1}{3} + \frac{2}{7}\right) \cdot x = \left(\frac{7}{21} + \frac{6}{21}\right) \cdot x = \mathbf{\frac{13}{21}x}$.

Example 17.11: Calculate the sum of $\frac{1}{4}$ of 60 and 6.

Solution: $\frac{1}{4} \cdot 60 + 6 = 15 + 6 = \mathbf{21}$.

Example 17.12: Calculate $\frac{1}{6}$ of a, where a is 72 decreased by 12.

Solution: $\frac{1}{6} \cdot a = \frac{1}{6}(72 - 12) = \frac{1}{6} \cdot 60 = \mathbf{10}$.

Example 17.13: $\frac{4}{7}$ of what number is equal to 16?

Solution: Let x be the number.

Then $\frac{4}{7} \cdot x = 16$. So $x = 16 \cdot \frac{7}{4} = 4 \cdot 7 = \mathbf{28}$.

Example 17.14: $\frac{3}{4}$ of the sum of a number and 3 is equal to 21. Find the number.

Solution: Let x be the number.

Then $\frac{3}{4}(x + 3) = 21$. So $x + 3 = 21 \cdot \frac{4}{3} = 7 \cdot 4 = 28$. So $x = \mathbf{25}$.

Example 17.15: Half of the sum of $\frac{1}{4}$ of a number and 6 is 24. Find the number.

Solution: Let x be the number.

Then $\frac{1}{2}\left(\frac{1}{4} \cdot x + 6\right) = 24$. So $\frac{1}{4} \cdot x + 6 = 24 \cdot 2 = 48$.

Therefore $\frac{1}{4}x = 48 - 6 = 42$, and so $x = 42 \cdot 4 = \mathbf{168}$.

ALGEBRA INVOLVING AGE COMPUTATIONS

Key expressions in age problems:

* An x year old person's age m years later is $x + m$.
* An x year old person's age m years earlier is $x - m$.
* The sum of the ages of four people in n years, whose ages now sum to m is $m + 4n$.
* The sum of the ages of four people n years earlier, whose ages now sum to m is $m - 4n$.
* The age difference of two siblings after n years, whose age difference now is m is still m.
* The average age of x people in n years, whose average age now is m is $m + n$
* The average age of x people n years earlier, whose average age now is m is $m - n$.

Example 17.16: Ahmet is now 12 years old and Mehmet is now 8 years old.

	Ahmet	Mehmet	Age difference	Sum of ages
Now	12	8	4	20
After 4 years	16	12	4	28
3 years ago	9	5	4	14
After m years	$12 + m$	$8 + m$	4	$20 + 2m$
n years ago	$12 - n$	$8 - n$	4	$20 - 2n$

Example 17.17: The ratio of Ahmet's to Mehmet's age is 1:3. If the sum of their ages is 20, what is their age difference?

Solution: Let x be Ahmet's age. It follows that Mehmet's age is $3x$, and $x + 3x = 20$. So $4x = 20$, and therefore $x = \frac{20}{4} = 5$.

Their age difference is $3x - x = 2x = 2 \cdot 5 = \mathbf{10}$.

Example 17.18: Ahmet is 20 years old and Mehmet is 16 years old. Calculate the ratio of Ahmet's age to Mehmet's age in 4 years.

Solution: In 4 years Ahmet will be 24 and Mehmet will be 20. So the ratio of Ahmet's age to Mehmet's age will be $\frac{24}{20} = \mathbf{\frac{6}{5}}$.

Example 17.19: The sum of the ages of 6 siblings is 85. Calculate the sum of their ages in 4 years.

Solution: The sum of their ages in 4 years is $85 + 4 \cdot 6 = 85 + 24 = \mathbf{109}$.

Example 17.20: The sum of the ages of 6 siblings is 85. Calculate the sum of their ages 3 years ago.

Solution: The sum of their ages 3 years ago: $85 - 3 \cdot 6 = 85 - 18 = \mathbf{67}$.

Example 17.21: Ahmet is 24, Mehmet is 12, and Veli is 6. When Mehmet's age becomes the same as Ahmet's age today, how old will Veli be?

Solution:

	Ahmet	Mehmet	Veli
Now	24	12	6
After 12 years	$24 + 12 = 36$	$12 + 12 = 24$	$6 + 12 = 18$

Veli will be **18**.

Example 17.22: If 6 years ago, the average age of 5 siblings was 20, then what will be their average age in 4 years?

Solution: 6 years ago the sum of the siblings' ages was $20 \cdot 5 = 100$. So in 4 years the sum of their ages will be $100 + 10 \cdot 5 = 150$.

So in 4 years their average age will be $\frac{150}{5} = \mathbf{30}$.

Example 17.23: A 40 year old father has a 12 year old daughter. In how many years will the ratio of their ages be 25:11 ?

Solution:

	Father	Daughter
	40	12
x years later	$40+x$	$12+x$

$$\frac{40+x}{12+x}=\frac{25}{11}\Rightarrow 11(40+x)=25(12+x)\Rightarrow 440+11x=300+25x$$

$$\Rightarrow 140=14x\Rightarrow x=\frac{140}{14}=\mathbf{10}.$$

Example 17.24: A 54 year old father has two children whose ages sum to 18. In how many years will the father's age be 2 times the sum of his children's ages?

Solution:

	Father	2 children
	54	18
x years later	$54+x$	$18+2x$

$$54+x=2(18+2x)\Rightarrow 54+x=36+4x\Rightarrow 18=3x\Rightarrow x=\frac{18}{3}=\mathbf{6}.$$

Example 17.25: A father's age is 22 more than the sum of his 2 children's ages. In 4 years the father's age will be 2 times the sum of his children's ages. How old is the father now?

Solution:

	Father	2 children
	$x+22$	x
In 4 years	$x+22+4$	$x+2\cdot 4$
	$x+26$	$x+8$

$x+26=2(x+8)=2x+16\Rightarrow x=26-16=10.$

The father's age is $x+22=10+22=\mathbf{32}$.

Example 17.26: The sum of the ages of Ahmet and Mehmet is 52. What will be the sum of their ages after 12 years?

Ahmet	Mehmet	Sum of ages
x	y	52 (now)
$x+12$	$y+12$	$x+y+12+12=52+24=\mathbf{76}$

Example 17.27: The sum of the ages of 60 students in a classroom is 600. Each student is either 8 or 12 years old. How many 8 year old students are there in the classroom?

Solution: Let x be the number of students who are 8 years old, and let y be the number of students who are 12 years old.

$$x + y = 60 \Rightarrow y = 60 - x$$

We also have

$$8x + 12y = 600$$

$$8x + 12(60 - x) = 600$$
$$8x + 720 - 12x = 600$$
$$-4x + 720 = 600$$
$$4x = 720 - 600 = 120$$
$$x = \frac{120}{4} = 30$$

So there are **30** students who are 8 years old.

WORK

Suppose that worker A can complete a job in m hours, worker B can complete the same job in n hours, and worker C can complete the job in k hours. Then we have the following:

- worker A can complete $\frac{1}{m}$ of the job in 1 hour.
- worker B can complete $\frac{1}{n}$ of the job in 1 hour.
- worker C can complete $\frac{1}{k}$ of the job in 1 hour.
- workers A and B can complete $(\frac{1}{m} + \frac{1}{n})t$ of the job in t hours.
- workers A, B, and C can complete $(\frac{1}{m} + \frac{1}{n} + \frac{1}{k})t$ of the job in t hours.

Example 17.28: If Ahmet can complete a job in 4 hours, then he can complete $\frac{1}{4}$ of the job in 1 hour. If Ahmet can complete a job in 20 days, then he can complete $\frac{1}{20}$ of the job in 1 day.

Example 17.29: Ahmet can complete a job in 10 days, and Mehmet can complete the same job in 15 days. What percent of the work can the two complete together in 1 day?

Solution: Ahmet can complete $\frac{1}{10}$ of the job in 1 day, and Mehmet can complete $\frac{1}{15}$ of the job in 1 day. So together they can complete

$$\frac{1}{10} + \frac{1}{15} = \frac{3+2}{30} = \frac{5}{30} = \frac{1}{6}$$

of the job in 1 day, and $\frac{1}{6} \cdot 100 = \frac{50}{3} = 16\frac{2}{3}$. So $\frac{1}{6} = \mathbf{16\frac{2}{3}}$%.

Example 17.30: Ahmet alone can complete a job in 6 days, and Mehmet alone can complete the same job in 8 days. If they work together, in how many days can they finish the job?

Solution: $\frac{1}{6}+\frac{1}{8}=\frac{1}{t}\Rightarrow\frac{4+3}{24}=\frac{1}{t}\Rightarrow\frac{7}{24}=\frac{1}{t}\Rightarrow t=\frac{24}{7}=\mathbf{3\frac{4}{7}}$ **days.**

Example 17.31: Ahmet alone can complete a job in 15 days, and Mehmet alone can complete the same job in 12 days. If they work together for 5 days, what percent of the job will be completed?

Solution: In 1 day, Ahmet can complete $\frac{1}{15}$ of the job, and Mehmet can complete $\frac{1}{12}$ of the job. $\left(\frac{1}{15}+\frac{1}{12}\right)\cdot 5=\frac{4+5}{60}\cdot 5=\frac{9}{12}=\frac{3}{4}=\mathbf{75\%}$.

Example 17.32: Ahmet can complete a job alone in 15 days, and Mehmet can complete the same job alone in 20 days. How many days would it take them to complete the job if they work together?

Solution: $\frac{1}{15}+\frac{1}{20}=\frac{1}{t}\Rightarrow\frac{4+3}{60}=\frac{1}{t}\Rightarrow\frac{7}{60}=\frac{1}{t}\Rightarrow t=\frac{60}{7}=\mathbf{8\frac{4}{7}}$ **days.**

Example 17.33: Ahmet can complete a job alone in 12 days, Mehmet can complete the same job alone in 10 days and Veli can complete the job alone in 30 days. If all three work together, in how many days can they complete the job?

Solution: $\frac{1}{12}+\frac{1}{10}+\frac{1}{30}=\frac{1}{t}\Rightarrow\frac{5+6+2}{60}=\frac{1}{t}\Rightarrow\frac{13}{60}=\frac{1}{t}\Rightarrow t=\frac{60}{13}=\mathbf{4\frac{8}{13}}$ **days.**

Example 17.34: Ahmet can complete a job alone in 20 days and Mehmet can complete the same job alone in 25 days. If Ahmet works for 5 days and Mehmet works for 10 days, what percent of the job will be completed?

Solution: In 1 day, Ahmet completes $\frac{1}{20}$ of the job and Mehmet completes $\frac{1}{25}$ of the job. $\frac{1}{20}\cdot 5+\frac{1}{25}\cdot 10=\frac{25}{100}+\frac{40}{100}=\frac{65}{100}=\mathbf{65\%}$.

Example 17.35: Ahmet can complete a job alone in 12 days, and Mehmet can complete the same job alone in 18 days. In how many days can they complete $\frac{7}{36}$ of the job if they work together?

Solution: In 1 day, Ahmet completes $\frac{1}{12}$ of the job and Mehmet completes $\frac{1}{18}$ of the job. We need to find t so that $\left(\frac{1}{12}+\frac{1}{18}\right)t=\frac{7}{36}$, or equivalently $\frac{3+2}{36}t=\frac{7}{36}$. So $5t=7$, and $t=\frac{7}{5}=\mathbf{1\frac{2}{5}}$ **days.**

Example 17.36: Ahmet can complete a job alone in 12 days and Mehmet can complete the same job alone in 8 days. After working together for 2 days Mehmet leaves the job. In how many days can Ahmet complete the rest of the job himself?

Solution: In 1 day, Ahmet completes $\frac{1}{12}$ of the job and Mehmet completes $\frac{1}{18}$ of the job. Let x be the number of days that Ahmet must work by himself after Mehmet leaves the job. Then

$2\left(\frac{1}{12}+\frac{1}{8}\right)+\frac{x}{12}=1$. So $\frac{1}{6}+\frac{1}{4}+\frac{x}{12}=1 \Rightarrow 2+3+x=12 \Rightarrow x=\mathbf{7}$.

Example 17.37: Ahmet can complete a job alone in 10 days, Mehmet can complete the same job alone in 15 days, and Veli can complete the job alone in 20 days. All 3 of them work together for 3 days and then Ahmet has to leave the job. In how many days can Mehmet and Veli complete the job together?

Solution:

$$3\left(\frac{1}{10}+\frac{1}{15}+\frac{1}{20}\right)+\frac{x}{15}+\frac{x}{20}=1 \Rightarrow 3(6+4+3)+4x+3x=60 \Rightarrow$$
$$39+7x=60 \Rightarrow 7x=60-39=21 \Rightarrow x=\frac{21}{7}=\mathbf{3}.$$

Example 17.38: One pipe can fill a pool in 10 hours, and a second pipe can fill the same pool in 30 hours. If both pipes are used, how many hours will it take to fill the pool?

Solution: $\frac{1}{10}+\frac{1}{30}=\frac{1}{t} \Rightarrow \frac{3+1}{30}=\frac{1}{t} \Rightarrow \frac{4}{30}=\frac{1}{t} \Rightarrow t=\frac{30}{4}=\mathbf{7\frac{1}{2}}$ **hours.**

Example 17.39: One pipe can fill a pool in 8 hours, and a second pipe can fill the same pool in 12 hours. A third pipe drains the pool in 16 hours. In how many hours will the pool be filled if all three pipes are used?

Solution: $\frac{1}{8}+\frac{1}{12}-\frac{1}{16}=\frac{1}{t} \Rightarrow \frac{6+4-3}{48}=\frac{1}{t} \Rightarrow \frac{7}{48}=\frac{1}{t} \Rightarrow t=\frac{48}{7}=\mathbf{6\frac{6}{7}}$ **hours.**

Example 17.40: Four pipes with the same capacity can fill $\frac{1}{6}$ of a pool in 8 hours. In how many hours can one of the pipes fill the empty pool?

Solution: If 4 pipes fill $\frac{1}{6}$ of the pool in 8 hours, then the 4 pipes can completely fill the pool in $6 \cdot 8 = 48$ hours. So 1 pipe can fill the pool in $48 \cdot 4 =$ **192 hours.**

Example 17.41: Two pipes with equal water flow capacity working together can fill a pool in 12 hours. If the water flow capacity of one of the pipes is increased by 6 times, and the water flow capacity of the other is reduced to $\frac{1}{6}$ of its original capacity, then how many hours will it take to fill the pool if both pipes are used?

Solution: Assume that each pipe can fill the pool in x hours working individually. We have $\frac{1}{x}+\frac{1}{x}=\frac{1}{12} \Rightarrow \frac{2}{x}=\frac{1}{12} \Rightarrow x=24$ hours. So each pipe working alone can fill the pool in 24 hours.

If the water flow capacity of the first pipe is increased by 6 times, then it will take $\frac{24}{6}=4$ hours to fill the pool. And if the flow capacity of the second pipe is reduced to $\frac{1}{6}$ of its original capacity, it will take $24 \cdot 6 = 144$ hours to fill the pool.

$\frac{1}{4}+\frac{1}{144}=\frac{1}{t} \Rightarrow \frac{36+1}{144}=\frac{1}{t} \Rightarrow \frac{37}{144}=\frac{1}{t} \Rightarrow t=\frac{144}{37}=3\frac{33}{37}$. So it will take the two pipes working together $\mathbf{3\frac{33}{37}}$ **hours** to fill the pool.

Example 17.42: One pipe can fill a pool in 8 hours and a second pipe can fill the same pool in 6 hours. If the two pipes are turned on when $\frac{1}{7}$ of the pool is full, how long will it take to fill the pool?

Solution: $\frac{1}{8}+\frac{1}{6}=\frac{1}{t} \Rightarrow \frac{3+4}{24}=\frac{1}{t} \Rightarrow \frac{7}{24}=\frac{1}{t} \Rightarrow t=\frac{24}{7}$. So both pipes together can fill the empty pool in $\frac{24}{7}$ hours.

Since $\frac{1}{7}$ of the pool was full, $\frac{6}{7}$ of the pool was empty. So it will take $\frac{6}{7} \cdot \frac{24}{7}=\frac{144}{49}=\mathbf{2\frac{46}{49}}$ **hours** to fill the rest of the pool.

Example 17.43: A pipe can fill an empty pool in 21 hours. If the amount of water flow is decreased by 30%, then how many hours will it take for the pipe to fill the pool?

Solution: $100 \cdot 21 = 70x \Rightarrow x=\frac{100 \cdot 21}{70}=10 \cdot 3=\mathbf{30}$ **hours**.

MIXTURES

When A grams of salt is mixed with B grams of water,

- The ratio of salt to water is $\frac{A}{B}$.
- The fraction of salt in the mixture is $\frac{A}{A+B}$ and the percentage of salt in the mixture is $\frac{A}{A+B} \cdot 100$.
- The fraction of water in the mixture is = $\frac{B}{A+B}$ and the percentage of water in the mixture is $\frac{B}{A+B} \cdot 100$.

Example 17.44: There are 16 liters of milk in container A and 32 liters of water in container B. Both liquids are mixed together into a single container. What is the percentage of milk in the mixture?

Solution: $\frac{16}{16+32} \cdot 100 = \frac{16}{48} \cdot 100 = \frac{100}{3} = \mathbf{33\frac{1}{3}}\%$.

Example 17.45: 40 liters of a 25% sugar solution is mixed with 30 liters of a 50% sugar solution. Find the percentage of sugar in the new solution.

Solution: $\frac{40 \cdot 0.25 + 30 \cdot 0.5}{40+30} \cdot 100 = \frac{1000+1500}{70} = \frac{2500}{70} = \mathbf{35\frac{5}{7}}\%$.

Example 17.46: 50 liters of water containing 40% salt is mixed with x liters of water containing 30% salt. If the new mixture contains 35% salt, then what is the value of x ?

Solution: $\frac{50 \cdot 0.4 + x \cdot 0.3}{50+x} \cdot 100 = 35 \Rightarrow \frac{2000+30x}{50+x} = \frac{35}{1}$

$$\Rightarrow 2000 + 30x = 1750 + 35x \Rightarrow 5x = 250 \Rightarrow x = \frac{250}{5} = \mathbf{50}.$$

Example 17.47: 600 grams of fruit juice containing 30% sugar is mixed with 800 grams of fruit juice containing 20% sugar. What is the percentage of sugar in the final mixture?

Solution: $\frac{600 \cdot 30 + 800 \cdot 20}{600+800} = \frac{18000+16000}{1400} = \frac{34000}{1400} = \mathbf{24\frac{2}{7}}\%$.

Example 17.48: When mixture A containing 10% salt is mixed with mixture B containing 20% salt, we get a 300 gram mixture containing 15% salt. How many grams did mixture A consist of?

Solution: Since 15 is the arithmetic mean of 10 and 20, mixtures A and B must be equal in quantity. So mixture A and B each have **150 grams**.

Example 17.49: When x grams of water is evaporated from 120 grams of water containing 20% salt, the salt percentage of the mixture increases to 40%. What is the value of x ?

Solution: $\frac{120 \cdot 0.2 + x \cdot 0}{120 - x} \cdot 100 = 40 \Rightarrow \frac{2400}{120 - x} = \frac{40}{1}$

$$\Rightarrow 2400 = 4800 - 40x \Rightarrow 40x = 2400 \Rightarrow x = \frac{2400}{40} = \mathbf{60}.$$

MOTION WORD PROBLEMS

distance = rate · time or $\boldsymbol{d = r \cdot t}$

Example 17.50: A car travels with an average speed of 60 km/h. What is the distance that the car travels in 6 hours?

Solution: $d = r \cdot t = 60 \cdot 6 =$ **360 km.**

Example 17.51: A vehicle travels with an average speed of 120 km/h. In how many hours will this vehicle travel a distance of 600 km?

Solution: $d = r \cdot t \Rightarrow 600 = 120t \Rightarrow t = \frac{600}{120} =$ **5 hours.**

Example 17.52: Running at a constant speed, an antelope traveled 200 miles in 5 hours. At this rate, how many miles did the antelope travel in 5 hours?

Solution: Using $d = r \cdot t$ we have $200 = r \cdot 5$, or $r = \frac{200}{5} = 40$ mph. Using $d = r \cdot t$ again, we have $d = 40 \cdot 4 =$ **160 miles.**

Example 17.53: Marco drove from home to work at an average speed of 50 miles per hour and returned home along the same route at an average speed of 46 miles per hour. If his total driving time for the trip was 4 hours, how many <u>minutes</u> did it take Marco to drive from work to home?

Solution: Let's put the given information into the following chart.

	Distance	Rate	Time
home to work	d	50	$\frac{d}{50}$
work to home	d	46	$\frac{d}{46}$
total			4

Note that although we do not know either distance, we do know that they are the same, so we can call them both "d." Also, since

$$\text{distance} = \text{rate} \cdot \text{time},$$

we have that time = $\frac{\text{distance}}{\text{rate}}$. We use this to get the first two entries in column three. The total time is given in the question. So we have

$$\frac{d}{50} + \frac{d}{46} = 4$$

$$46d + 50d = 4 \cdot 50 \cdot 46$$

$$96d = 4 \cdot 50 \cdot 46$$

$$d = \frac{4 \cdot 50 \cdot 46}{96}$$

We want the time it takes Marco to drive from work to home, that is we want $\frac{d}{46}$.

This is equal to $\frac{d}{46} = \frac{4 \cdot 50}{96}$ in hours. To convert to minutes we multiply by 60.

$\frac{d}{46} = \frac{4 \cdot 50 \cdot 60}{96} =$ **125 minutes.**

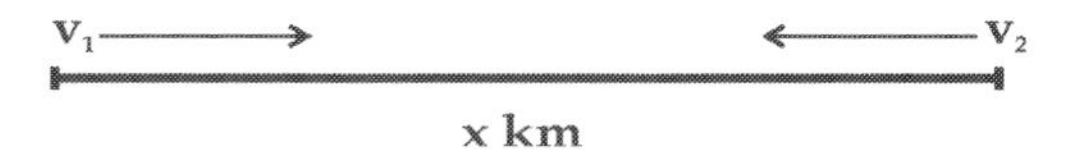

When two cars with a distance of x between them move towards each other with speeds of v_1 and v_2, respectively, they will collide at time $t = \frac{x}{v_1+v_2}$.

Notes: (1) Here we are using the same formula $d = r \cdot t$, where $d = x$ and $r = v_1 + v_2$. So we have $x = (v_1 + v_2)t$, or equivalently $t = \frac{x}{v_1+v_2}$.

(2) The same formula can be used if two cars starting at the same position travel in opposite directions with speeds of v_1 and v_2, respectively, along a circular track with circumference x.

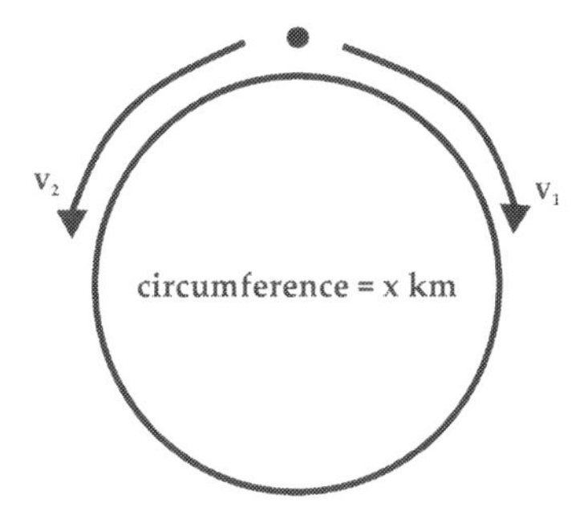

Example 17.54: There is a distance of 1200 km between two cars. The cars move towards each other with speeds of 80 km/h and 40 km/h, respectively. In how many hours will the cars collide?

Solution: $t = \frac{1200}{80+40} = \frac{1200}{120} =$ **10 hours**.

V1 → V2 →

x km

When two cars with a distance of x between them move in the same direction with speeds $v_1 > v_2$, then the rear car will overtake the forward car at time $t = \frac{x}{v_1-v_2}$.

Notes: (1) Here we are using the same formula $d = r \cdot t$, where $d = x$ and $r = v_1 - v_2$. So we have $x = (v_1 - v_2)t$, or equivalently $t = \frac{x}{v_1-v_2}$.

(2) The same formula can be used if two cars starting at the same position travel in the same direction with speeds of v_1 and v_2, respectively, along a circular track with circumference x.

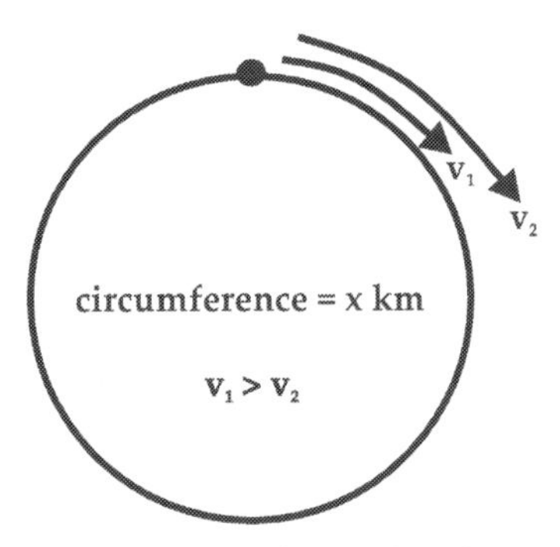

Example 17.55: There is a distance of 1200 km between two cars. The cars move in the same direction with speeds of 80 km/h and 40 km/h, respectively. In how many hours will the rear car overtake the forward car?

Solution: $t = \frac{1200}{80-40} = \frac{1200}{40} = \mathbf{30}$ **hours.**

The **Harmonic Mean Formula** can be used to find an average speed when two individual speeds for the same distance are known:

$$\text{Average Speed} = \frac{2(\text{Speed 1})(\text{Speed 2})}{\text{Speed 1} + \text{Speed 2}} \quad \text{or} \quad v_{avg} = \frac{2v_1v_2}{v_1+v_2}$$

Note: This formula is equivalent to the Harmonic Mean Formula given in chapter 6. There we were given $v_{avg} = \frac{2}{\frac{1}{v_1}+\frac{1}{v_2}}$.

We get the equivalent version above by multiplying the numerator and denominator of this expression by v_1v_2.

Example 17.56: Jason ran a race of 1600 meters in two laps of equal distance. His average speeds for the first and second laps were 11 meters per second and 7 meters per second, respectively. What was his average speed for the entire race, in meters per second?

Solution: $v_{avg} = \frac{2(11)(7)}{11+7} = \frac{154}{18} = \frac{\mathbf{77}}{\mathbf{9}}$ **meters/sec.**

Example 17.57: Two cars with speeds of 60 km/h and 40 km/h, respectively, move from city A to city B starting at the same point at the same time. The faster car arrives at city B 4 hours earlier than the slower car. What is the distance from city A to city B?

Solution: $d = 60t, d = 40(t+4) \Rightarrow 60t = 40(t+4) = 40t + 160$

$$\Rightarrow 20t = 160 \Rightarrow t = 8 \Rightarrow d = 60t = 60 \cdot 8 = \mathbf{480\ km}.$$

18. OTHER BASES

Recall that in the decimal numeration system there are ten symbols 0,1,2,3,4,5,6,7,8 and 9. We must therefore group objects by tens.

Example 18.1: Draw eleven squares and group them 10 at a time. Write the number of squares in base 10.

Solution:

Notice that we are grouping 10 of the squares together. So we have 1 group of squares and 1 extra square. So we write this number as 11.

In base five there are five symbols 0,1,2,3 and 4. We therefore group by fives.

Example 18.2: Draw eleven squares and group them 5 at a time. Write the number of squares in base 5.

Solution:

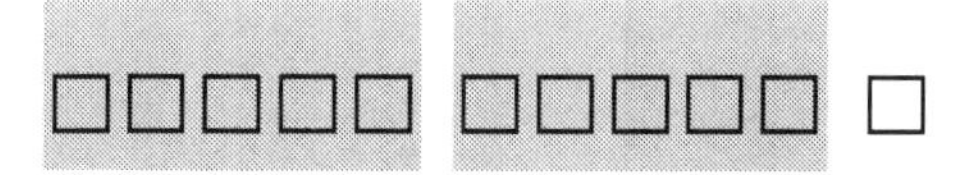

Notice that we have 2 groups of 5 squares and 1 extra square. So we write this number as 21_{five}.

Example 18.3: Write the number 11 in base 8 and base 4.

Solution:

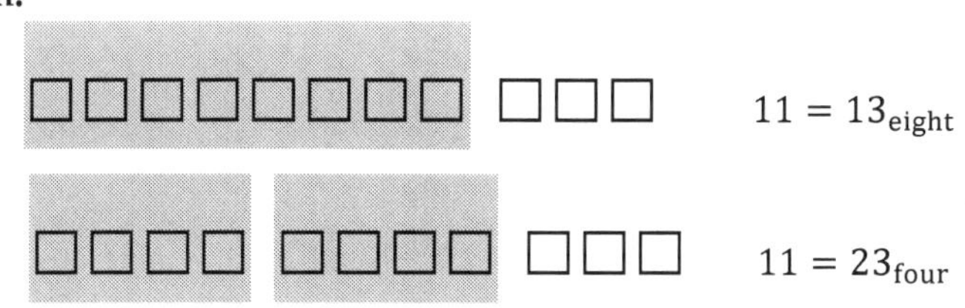

Example 18.4: Write the number 11 in base 3 and base 2.

Solution:

Notice that we have 3 groups of 3 squares. But in base three there is no symbol for 3 groups (the symbols are 0, 1, and 2). So we group those 3 groups together into 1 "supergroup" (which we displayed with the underline). So we have 1 supergroup, 0 groups, and 2 extra squares.

$11 = 1011_{\text{two}}$

For bases higher than 10 we can begin using the letters A, B, C,...

For example, in base eleven there are 11 symbols: 0,1,2,3,4,5,6,7,8,9,A.

In base twelve there are 12 symbols: 0,1,2,3,4,5,6,7,8,9,A,B.

Example 18.4: Write the number 11 in base 11 and base 12.

Solution: $11 = 10_{\text{eleven}}$ and $11 = A_{\text{twelve}}$

CHANGING TO BASE TEN

The expanded form of a base ten number shows both the **place values** and **face values** of the number. For example, the expanded form of 2367 is $2 \cdot 10^3 + 3 \cdot 10^2 + 6 \cdot 10^1 + 7 \cdot 10^0$. The face values of 2, 3, 6, and 7 correspond to the place values of 10^3, 10^2, $10^1 = 10$, and $10^0 = 1$.

If we want to change a number from a different base to base ten, we simply write the expanded form of the number where the place values are powers of the given base, and the face values are the base ten equivalents of the digits in each position.

Example 18.5: Change 65_{eight} to base ten.

Solution: $6 \cdot 8^1 + 5 \cdot 8^0 = 6 \cdot 8 + 5 = 48 + 5 = \mathbf{53.}$

Example 18.6: Change 4312_{five} to base ten.

Solution: $4 \cdot 5^3 + 3 \cdot 5^2 + 1 \cdot 5^1 + 2 \cdot 5^0 = 4 \cdot 125 + 3 \cdot 25 + 1 \cdot 5 + 2$

$$= 500 + 75 + 5 + 2 = \mathbf{582.}$$

Example 18.7: Change 212000_{three} to base ten.

Solution: $2 \cdot 3^5 + 1 \cdot 3^4 + 2 \cdot 3^3 = 2 \cdot 243 + 1 \cdot 81 + 2 \cdot 27$

$$= 486 + 81 + 54 = \mathbf{621.}$$

Example 18.7: Change $52A_{\text{twelve}}$ to base ten.

Solution: $5 \cdot 12^2 + 2 \cdot 12^1 + 10 \cdot 12^0 = 5 \cdot 144 + 2 \cdot 12 + 10$

$$= 720 + 24 + 10 = \mathbf{754.}$$

CHANGING FROM BASE TEN TO ANOTHER BASE

To convert a number from base ten to another base, we divide the number by the given base, keeping track of the remainder. We then proceed to divide each quotient by the given base, again keeping track of remainders. The digits of the number from left to right are the remainders written from last to first.

Example 18.8: Convert 435 into base six.

Solution:

$$435 \div 6 = 72\ R3$$
$$72 \div 6 = 12\ R0$$
$$12 \div 6 = 2\ R0$$
$$2 \div 6 = 0\ R2$$

So $435 = \mathbf{2003_{six}}$.

Example 18.9: Convert 23 into base five.

Solution:

$$23 \div 5 = 4\ R3$$
$$4 \div 5 = 0\ R4$$

So $23 = \mathbf{43_{five}}$.

Example 18.10: Convert 75 into base two.

Solution:

$$75 \div 2 = 37\ R1$$
$$37 \div 2 = 18\ R1$$
$$18 \div 2 = 9\ R0$$
$$9 \div 2 = 4\ R1$$
$$4 \div 2 = 2\ R0$$
$$2 \div 2 = 1\ R0$$
$$1 \div 2 = 0\ R1$$

So $75 = \mathbf{1001011_{two}}$.

Example 18.11: Convert 5133 into base twelve.

Solution:

$$5133 \div 12 = 427\ R9$$
$$427 \div 12 = 35\ R7$$
$$35 \div 12 = 2\ R11$$
$$2 \div 12 = 0\ R2$$

So $5133 = \mathbf{2B79_{twelve}}$.

Example 18.12: Convert 172 into base fifteen.

Solution:

$$172 \div 15 = 11\ R7$$
$$11 \div 15 = 0\ R11$$

So $172 = \mathbf{B7_{fifteen}}$.

ARITHMETIC IN OTHER BASES

The same algorithms that we already know for addition, subtraction, and multiplication can be used in other bases. We just need to perform each individual computation in the given base. For example, when adding 1 and 1 in base two, we get 10_{two}. So we would write down the 0 and carry the 1.

Example 18.13: Let $x = 11011_{two}$ and $y = 10010_{two}$. Compute $x + y$, $x - y$, and $x \cdot y$.

Solution:

```
          1
x + y:  11011        x - y:  11011        x · y:     11011
      + 10010              - 10010                 × 10010
        101101                1001                   00000
                                                    110110
                                                   0000000
                                                  00000000
                                                 110110000
                                                 111100110
```

Notes: (1) For addition, we begin at the right. Since $1 + 0 = 1$, we place a **1** in the rightmost position. Moving to the left we have $1 + 1 = \mathbf{10}$. So we place a **0** in the next position to the left and carry the 1. We then have $1 + 0 + 0 = \mathbf{1}$, $1 + 0 = \mathbf{1}$, and $1 + 1 = \mathbf{10}$.

(2) For subtraction, we again begin at the right. Since $1 - 0 = 1$, we place a **1** in the rightmost position. Moving to the left we have $1 - 1 = \mathbf{0}$, $0 - 0 = \mathbf{0}$, $1 - 0 = \mathbf{0}$, and $1 - 1 = 0$ (we do not write down this last 0).

(3) For multiplication we begin by multiplying the rightmost 0 by each of the digits in 11011 moving from right to left to get 00000. We then move one digit to the left, put down a ***0*** as a placeholder, and then multiply 1 by each of the digits in 11011 moving from right to left to get 11011. We continue in this fashion until we have taken care of all of the digits in 10010. We then add all of these rows using the algorithm for addition.

(4) When multiplying it's more convenient to skip row of zeros. Just remember to put in the proper number of placeholders. The simplified computation looks like this:

$x \cdot y$: 11011

× 10010

11011***0***

11011***0000***

111100110

(4) As an alternative to performing the operations in base two, we can change each of the numbers to base ten, perform the operation in base ten, and then change back to base two.

$$11011 = 1 \cdot 2^4 + 1 \cdot 2^3 + 1 \cdot 2 + 1 = 16 + 8 + 2 + 1 = 27$$

$$10010 = 1 \cdot 2^4 + 1 \cdot 2 = 16 + 2 = 18$$

So for example, for addition we have $27 + 18 = 45 = \mathbf{101101_{two}}$.

Example 18.14: Let $x = 2020_{\text{three}}$ and $y = 1200_{\text{three}}$. Compute $x + y$, $x - y$, and $x \cdot y$.

$x + y$: 2020

\+ 1200

10220

$x - y$:

1

$2^{1}020$

− 1 200

120

$x \cdot y$:

1

2020

× 1200

11110***00***

2020***000***

10201000

Example 18.15: Add $x = 42\text{F}_{\text{sixteen}}$ and $y = \text{B}3_{\text{sixteen}}$.

Solution:

```
   1
  42F
+  B3
  4E2
```

Example 18.15: Subtract $y = 2A_{\text{twelve}}$ from $x = B1_{\text{twelve}}$.

Solution:

```
  A
  B¹1
- 2A
   82
```

Example 18.15: Multiply $x = 415_{\text{eleven}}$ and $y = 200_{\text{eleven}}$.

Solution:

```
   415
×  200
 82A00
```

Additional Problems Questions

Full solutions to these problems are available for free download here:

www.thesatmathprep.com/McountX.html

Numbers

1. What is the result when the numbers 764 and 587 are each rounded to the tens place and then added together?
2. $\frac{3}{7}$ is what percent of $\frac{4}{21}$?
3. If 30 liters of water is poured into 0.8 liter bottles, how many bottles would be completely filled?
4. If an item with a selling price of $120 is discounted 20%, what is the discounted price?
5. $8 \cdot 6^3 + 6 \cdot 6^3 + 12 \cdot 6^3 + 10 \cdot 6^3 =$
6. 4.375 can be written as the mixed numeral $a\frac{b}{c}$ where GCD(b,c) = 1. Evaluate $a + b + c$.
7. An increase of 11% followed by a decrease of 17% is equivalent to a single decrease of what percent? Round your answer to the nearest tenth of a percent.
8. In a basketball tournament, a team lost $\frac{1}{3}$ of its matches and tied $\frac{1}{4}$ of its matches. If this team won 15 matches, how many matches were tied in all?
9. If 40% of a is equal to 5% of b, what percent of a is b?
10. If a and b are positive integers such that 40 times a is equal to 50 times b, then what is the minimum value of $a + b$?
11. Change $2C5_{\text{fourteen}}$ to base ten.
12. What is the greatest whole number that is a factor of both 2430 and 2431?

13. If today is Sunday, then what day of the week will it be in 150 days?

14. If 4 is added to the numerator of the fraction $\frac{5}{8}$, by what percent would the fraction increase?

15. What is the sum of the distinct prime factors of 798?

16. A purchase of 28 kg of sugar is made. If there is a 30% sale, how many kg of sugar can be purchased for the same amount of money?

17. Evaluate $\frac{6\cdot6!-5!}{4\cdot4!-3!}$

18. Find the sum of the positive integers a for which $\frac{2a+20}{a}$ is a positive integer.

19. What is the absolute difference between 2.13×10^5 and 3.57×10^3 ?

20. Find the sum of the positive divisors of 50.

21. If $\frac{(n+1)!}{(n-1)!} = 56$, find n.

22. The LCM of two positive integers is 144 and the GCD of the same two integers is 2. What is the least possible value for the sum of the two integers?

$$-\frac{7}{4} < x < \frac{11}{3}$$

23. What is the sum of all integers x satisfying the above inequality?

24. What common fraction is equivalent to $0.\overline{405}$?

25. If the sum of the LCM and GCD of two consecutive positive integers is 133, what is the sum of the two integers?

26. If a and b are positive integers such that $5a + 7b = 700$, then what is the maximum possible value of b ?

27. If today is Friday, then what day of the week will it be in 324 days?

28. Let a and b be positive integers satisfying $\frac{a}{5} + 2b = 9$. What is the maximum possible value of a?

29. What is the least positive integer k such that the product $275k$ is a perfect square?

30. Add $x = 1322_{\text{four}}$ and $y = 3132_{\text{four}}$.

31. Consider an arithmetic sequence with $a_7 = 312$ and $a_{12} = 567$. For what value of n is $a_n = 2760$?

32. If $k! = \frac{(7!)!}{7!}$, what is the value of k ?

33. The product of two nonnegative integers is 1,000,000. If neither number is a multiple of ten, what is the absolute value of the difference between these two integers?

$$0 < x < 1,\ a = \frac{1}{x},\ b = x^2,\ c = \sqrt{x}$$

34. According to the information above, put a, b, and c in order from greatest to least.

35. If a and b are positive integers with $-5 < a < 0 < b < 7$, find all possible values of $a + b$.

36. How many positive integers greater than 100 and less than 400 give a remainder of 3 when divided by 5?

37. What is the sum of the first 100 terms of the arithmetic sequence 3, 15, 27,...?

38. How many people must be in a group to guarantee that at least four people in the group were born on the same day of the week?

39. $\frac{3^n+3^n+3^n+3^n}{2^n+2^n+2^n+2^n} =$

40. How many digits does $16^4 \cdot 5^{10}$ have?

41. If $\sqrt{3} + \sqrt{12} + \sqrt{27} + \sqrt{48} + \sqrt{75} = a\sqrt{b}$, then $a + b =$

42. At an airport, beginning at 8:00 AM, a plane takes off from runway A every 20 minutes, a plane takes off from runway B every half hour, and a plane takes off from runway C every 15 minutes. When is the first time all three runways have a plane taking off at the same time?

43. Convert 357 into base eleven.

44. An arithmetic sequence has first term a and common difference d. If the sum of the first fifteen terms is one third of the sum of the next twenty terms, what is the ratio $\frac{d}{a}$? Express your answer as a common fraction.
45. What is the sum of the positive divisors of 2187?
46. If the sum of the first 26 positive odd integers is reduced by 9, what is the sum of the prime factors of the result?
47. If $a : b = 2 : 7$ and $b : c = 3 : 4$, what is the value of $\frac{c}{a}$? Express your answer as a common fraction.
48. Multiply $x = 236_{\text{seven}}$ and $y = 120_{\text{seven}}$.
49. For some real number a, the first three terms of an arithmetic sequence are $6a - 3$, $8a + 3$, and $9a$. What is the fourth term of the sequence?
50. Ken's wallet contains d dollars, where d is a two digit multiple of 13 that gives a remainder of 1 when divided by 3, 5, 9, and 15. If Ken's wallet has seven times the amount of money that Dan's wallet has, how much money is in Dan's wallet?

ALGEBRA AND FUNCTIONS

51. If a bookseller sells his books for $4 each he will make a $100 profit, and if he sells his books for $3 each he will take an $80 loss. How many books does the bookseller have?
52. There are 28 houses on a street. Each house has either 3 rooms or 4 rooms. If there are a total of 100 rooms, how many houses have 4 rooms?
53. If $x > y > 1$ and $(xy)^k = (x^2y^{-5})(x^{-4}y^3)$, what is k ?
54. 32 men and 8 women are in a room when several married couples enter. There are now three times as many men as women. How many couples entered the room?

$$a + b = 14$$
$$a + c = 12$$
$$b + c = 10$$

55. In the system of equations above, find a, b, and c.

56. If x and y are nonzero real numbers such that $\frac{6x+3y}{x} = 7$, then find $\frac{3y+x^2}{y}$ in terms of y.

57. If $7x$ is subtracted from 77 and the difference is then divided by 3x, the result is 14. What is the value of x ?

58. If $\frac{a}{b} = \frac{c}{d} = \frac{e}{f} = k$, find $\frac{a+b}{b} + \frac{d+c}{d} + \frac{f-e}{f}$ in terms of k.

59. Jessica can finish a job in 8 days and Rita can finish the same job in 12 days. If the two of them work together, in how many days will they complete the job?

60. A salesman sells products A, B and C for a total of $320. If one fifth of product A's price is equal to one fourth of product B's price and one seventh of product C's price, then find the price of product C.

61. A 30 year old woman has an 8 year old daughter. In how many years will the sum of their ages be 58?

62. If $\frac{a}{b} = \frac{1}{8}$, find $\left(\frac{a-b}{b}\right)^2$.

63. One pack of bird feed lasts 80 days for a canary, 40 days for a pigeon, and 20 days for a chicken. How many days will one pack of bird feed last if it is used for all 3 birds?

64. If Joseph were born 3 years earlier, his age would now be x and if he were born 5 years later, his age would be y. Today Joseph's age is 20. Find $x + y$.

65. If Jennifer were to give $\frac{4}{13}$ of her money to Elaine, they would have the same amount of money. Find the ratio of the amount of money Jennifer has to the amount of money Elaine has.

66. Let a, b, and c be positive integers with $ab = 36$, $bc = 144$, and $ac = 64$. Find abc.

67. If $\frac{a}{b} = \frac{1}{7}$, evaluate $\frac{a+b}{b-a}$.

68. If $3^n + 2^n = m$ and $3^n - 2^n = k$, express $3^{2n} - 2^{2n}$ in terms of m and k.

69. Ahmet works 6 times as fast as Mehmet. If Ahmet and Mehmet can complete a job together in 24 days, then how many days does it take for Mehmet to finish the job by himself?

70. If $a^x b^y c^z d^w = (a^3 b^5 c^{-1} d^2)^5$, find $x + y + z + w$.

71. What is the sum of the real values of x satisfying the equation $x + 3|x| - 5 = 0$?

72. Given the function $f(x) = 3x^2 - 12x + 9$ with domain $-1 \leq x \leq 4$, what is the largest value in the range of f ?

73. A bucket is partially filled with water. If 120 liters of water is added to the bucket, the bucket will contain a liters of water. If 80 liters of water is removed from the bucket, the bucket will contain $\frac{a}{4}$ liters of water. How many liters of water are in the bucket?

74. Let x be a positive integer such that $\frac{x+6}{2} > 12$. What is the least possible value for x ?

75. If $\frac{a}{3} + \frac{b}{5} + \frac{c}{7} = 12$, then evaluate $35a + 21b + 15c$.

76. If $x = 5y - 8$ and $y = z^2$, what is the smallest possible value of z for which $x = y$?

77. If $x^2 + 81 = -18x$, what is the value $\frac{x}{18} + 3$?

78. 2 vehicles moving at 6 meters/second and 4 meters/second travel around a circle with circumference 90 meters. The vehicles move in opposite directions starting from the same position. How many seconds later will they first meet?

79. If $x \neq 0$ and $y \neq 0$, then what is the simplest form of $\frac{\frac{x}{y} - \frac{y}{x}}{\frac{y}{x} + 1}$? Express your answer as a common fraction in terms of x and y.

80. If $2^n \cdot 3^n = k$, express $4^n \cdot 9^n$ in terms of k.

81. How much more is $6a - 3b$ than $3b - 6a$?

82. Given that $5 = \frac{m}{40} = \frac{70}{n}$, what is the value of $m + n$?

83. The sum of three numbers is 131. The first number is 8 more than the second number, and the second number is 6 less than the third number. Find the third number.

84. If $\frac{a+b}{b} = 7$, find $\frac{a-b}{b} + \frac{b-a}{a}$.

85. If $\frac{a}{b} = \frac{c}{d} = \frac{e}{f} = 6$, then find $\frac{a+b}{b} \cdot \frac{c+d}{d} \cdot \frac{e+f}{f}$.

86. If ⊠ + ⊡= 7 and ⊠ − ⊡= 3, what is the value of ⊡ + ⊡ ?

87. The ratio of Toni's age 6 years ago to her age 6 years from now is 1:4 How old is Toni today?

88. The price for 6 notebooks and 2 pencils is $20. The price for 3 notebooks and 7 pencils is $16. What is the price for 1 notebook and 1 pencil?

89. An elephant traveled 7 miles at an average rate of 4 miles per hour and then traveled the next 7 miles at an average rate of 1 mile per hour. What was the average speed, in miles per hour, of the elephant for the 14 miles?

90. If $g(x) = 3x^2 + 17$, what is the positive value of x for which $g(x) = 164$?

91. A ball thrown upward from a height of 7 feet with an initial velocity of v feet per second will be $-16t^2 + vt + 7$ feet above the ground after t seconds. What is the minimum value of v for which the ball reaches a height of at least 23 feet?

92. Define $f^1(x) = f(x)$ and $f^n(x) = f(f^{n-1}(x))$ for $n > 1$. For example, $f^2(x) = f(f(x))$. If $f^2(x) = 3x^2 - 1$, compute $f^6(1)$.

93. A teacher spends one fifth of his salary on rent, one twelfth of his salary on utilities, and one fourth of his salary on other expenses. If the remaining salary is $2800 each month, what is the annual salary of this teacher?

94. If $5^x - 1 = 7$, what is the value of $25^x - 3$?

95. If a vehicle were to travel 90 kph from location A to location B, the driver would arrive 10 minutes late. If the same vehicle were to travel 110 kph the driver would arrive 30 minutes early. Assuming that the driver arrives on time, how long does it take the vehicle to travel from location A to location B?

96. If $f(x) = 2x - 1$ and $g(x) = 5x + 3$, what is the value of $g(f(-2))$?

97. Solve for x: $5^{(2^x)} = 25^{(8^{12})}$.

98. A 20 liter solution that contains 10% sugar and an 80 liter solution that contains 20% sugar are mixed together. What is the percentage of sugar in the final mixture?

99. If $\frac{1}{x}+\frac{1}{y}+\frac{1}{z}=7$ and $\frac{1}{xyz}=13$, find the average of xy, xz, and yz.

100. Jen and Rachel can finish a job in 10 days. Rachel and Denise can finish the same job in 15 days. Jen and Denise can finish the job in 20 days. How many days would it take for Jen to finish the job by herself?

PROBABILITY, STATISTICS, AND DATA

101. The average (arithmetic mean) of three numbers is 57. If one of the numbers is 16, what is the sum of the other two?

102. The table below shows the amount of books sold in the years 2006, 2007, and 2008.

	2006	2007	2008
English	3600	4000	4500
History	4000	6000	3000
Biology	1200	4000	2000
Math	2200	7000	8000

What percent of the history books sold in the three years was sold in 2007?

103. All the employees at a certain company are Spanish, German, or French. The table below shows the number of employees of each nationality, classified by gender.

	Spanish	**German**	**French**	**Total**
Male	80	20	15	115
Female	120	30	5	155
Total	200	50	20	270

What is the probability that a randomly chosen employee from this company will be French or female?

104. From 6 rabbits and 7 chickens, in how many different ways can 4 rabbits and 3 chickens be chosen?

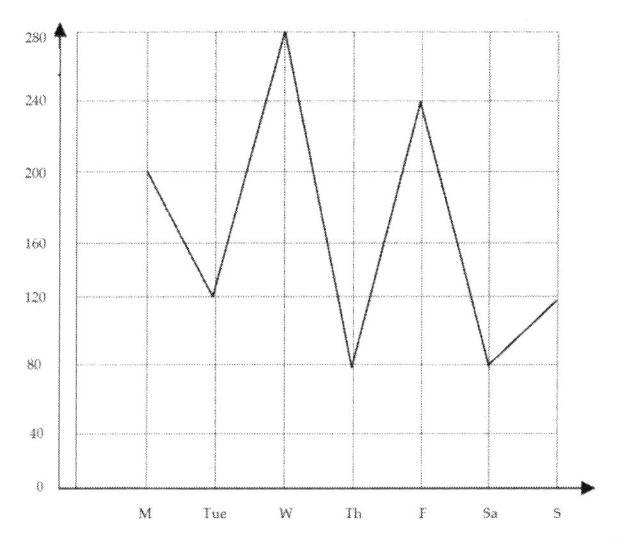

105. The above graph shows the number of passengers that were transported by a bus company during a specific week. How many passengers did the company transport during that week?

106. In a class of 48 students, each student plays either football or basketball or both. If 6 students play both football and basketball and 16 students play only basketball, how many students play only football?

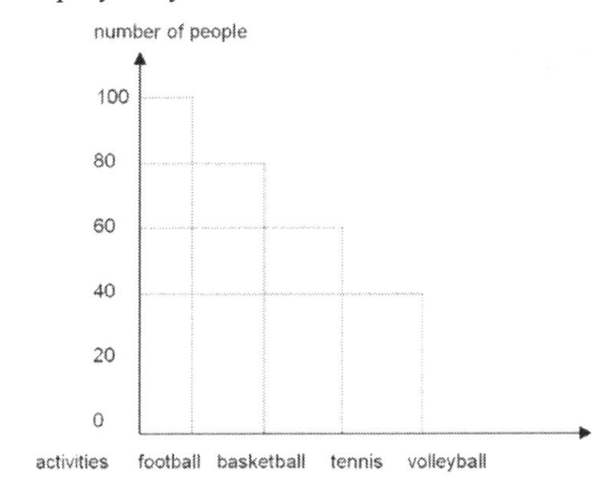

107. The price of one football ticket is $6 and the price of one basketball ticket is $8. Using the graph above, determine the difference between the money earned for football tickets and the money earned for basketball tickets.

108. A group of 4 people is formed from 6 men and 4 women. In how many ways can this group be chosen without any gender restrictions?

Book	Price
Math	90
Physics	85
Chemistry	140
Biology	160

109. The chart above shows a bookstore's prices for some of their textbooks. Maria bought one book from one subject and five books from a different subject. If she paid $565, which books did she buy?

110. Find the arithmetic mean of $\frac{1}{2}, \frac{3}{4}, \frac{5}{8},$ and $\frac{7}{16}$.

111. A group of 4 people is formed from 6 men and 4 women. In how many ways can this group be chosen so that the group contains no women?

112. In a group of 40 students, 32 are members of the music club and 20 are members of the art club. How many students are members of both clubs?

113. How many unique license plates with three letters followed by 3 numbers can be made if each of the symbols P, Q, R, 7, 8, and 9 are used exactly once?

114. 4 teachers and 6 students sit around a round table. In how many different ways can seats be chosen?

115. If the set A has 512 subsets, how many elements are in the set A ?

116. There are 12 boys and 14 girls in a class. How many different 5 student groups can be formed which include 2 boys and 3 girls?

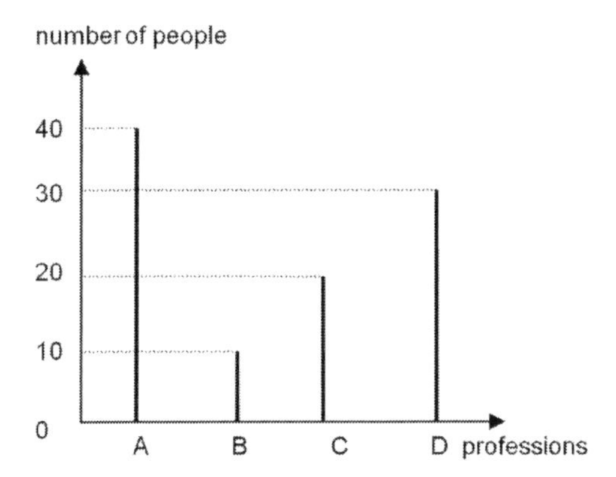

(A: teachers, B: doctors, C: crafts, D: technicians)

117. The graph above shows the number of people in a small town that work in the four given professions. When this information is displayed as a circle graph, what is the angle measure of the sector displaying the number of doctors?

118. Suppose that tomorrow there is a 50% chance of rain in New York and a 20% chance of rain in Tokyo. Assuming that these events are independent, what is the probability that it will rain in at least one of these cities?

119. A basket contains 16 balls of the same size numbered from 1 to 16. Two balls are randomly selected, one at a time from the basket. The first is replaced before the second ball is chosen. What is the probability that the first ball is numbered with an even number, and the second ball is numbered with an odd number?

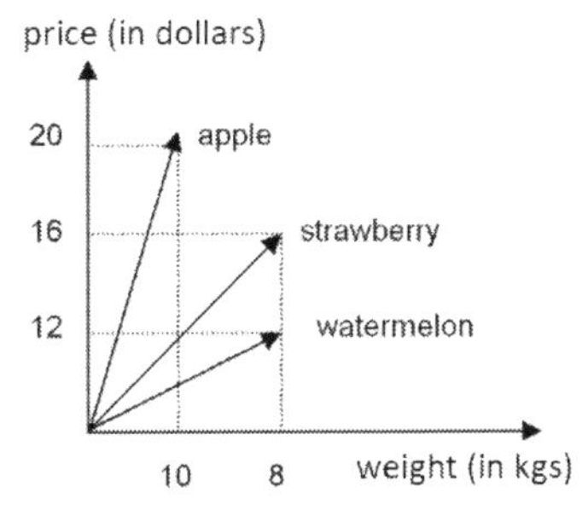

120. The above graph shows the sale price of apples, strawberries and watermelons in a supermarket. According to the graph, if a purchase is made for 5 kg of apples, 12 kg of strawberries and 2 kg of watermelon, what will be the total cost?

121. The mean and median of a list of eight integers are both 4.5. The smallest integer is 2, the largest integer is 6, and the only mode is 4. How many of the integers in the list are equal to 4?

122. The arithmetic mean of three consecutive positive integers is 42. What is the least of the three integers?

123. Six different 3 digit numbers can be created using the numbers 2, 4, and 6. Find the sum of these six numbers.

124. The mean of a set of k numbers is 53, and the mean of another set of $4k$ numbers is 72. What is the mean of a third set of $5k$ numbers if the mean of all $10k$ numbers is 85?

125. In a basket there are 9 balls numbered from 1 to 9. If 2 balls are picked from the basket one after another without returning the first ball to the basket, what is the probability both balls will be prime numbers?

126. $\frac{P(6,3)}{P(6,2)} =$

127. A class has a total of 40 students, 24 of which are female. 10 of the female students are blonde and 15 of the male students are blonde. If a student is chosen at random, what is the probability that the student will be either a female blonde or a male that is not blonde?

128. A basket contains 12 white balls, 7 green balls, 4 red balls, 5 yellow balls, and 2 black balls. If two balls are randomly selected from the basket, what is the probability that they will be different colors?

129. The arithmetic mean of 14 numbers is 22. If 4 of the numbers with average 12 are eliminated, what is the arithmetic mean of the remaining 10 numbers?

130. 5 teams are playing in a soccer tournament. How many games will be played if each team plays 4 games against every other team?

2	4 4 5 7
3	1 2 6
4	7 7 8
5	5 9
6	1 2 3 4 5

131. In the stem-and-leaf-plot above, the line 2| 4 4 5 7 represents the data 24, 24, 25, 27. Find the median of the data in the stem-and-leaf-plot.

132. The mean of five whole numbers is 13, and the unique mode is 14. If all five numbers are greater than 10 but less than 15, what is the median of the five numbers?

133. Jason answered 95% of the first part of a quiz correctly. He then correctly answered 5 of the remaining 20 questions. How many questions were on the quiz if Jason wound up with an average of 55%?

134. The integers 1 through 15 are written in base three, and a digit is chosen at random What is the probability that the digit chosen is 2?

135. 100 marbles labeled with consecutive integers are in a box. Two are drawn at random and without replacement. What is the probability that the absolute difference between the two numbers is less than 16? Express your answer as a common fraction.

ANSWERS TO ADDITIONAL PROBLEMS

1. 1350
2. 225
3. 37
4. $96
5. 7776
6. 15
7. 7.9%
8. 9
9. 800%
10. 9
11. 565
12. 1
13. Wednesday
14. 80%
15. 31
16. 40 kg
17. $\frac{140}{3}$
18. 42
19. 209,430
20. 93
21. 7
22. 34
23. 5
24. $\frac{15}{37}$
25. 23
26. 95
27. Sunday
28. 35
29. 11
30. 11120_{four}
31. 55
32. 5039
33. 15,561
34. $a > c > b$
35. -3, -2, -1, 0, 1, 2, 3, 4, 5
36. 60
37. 59,700
38. 22
39. $\left(\frac{3}{2}\right)^n$
40. 12
41. 18
42. 9:00 AM
43. $2\text{A}5_{\text{eleven}}$
44. $\frac{1}{7}$
45. 3280
46. 52
47. $\frac{14}{3}$
48. 31650_{seven}
49. -93
50. 13 dollars
51. 180
52. 16
53. -2
54. 4
55. $a = 8, b = 6, c = 4$
56. $3 + 9y$
57. $\frac{11}{7}$
58. $k + 3$
59. $4\frac{4}{5}$ days
60. $140
61. 10 years
62. $\frac{49}{64}$
63. $11\frac{3}{7}$ days

64. 38
65. $\frac{13}{5}$
66. 576
67. $\frac{4}{3}$
68. mk
69. 168
70. 45
71. $-\frac{5}{4}$ or -1.25
72. 24
73. $146\frac{2}{3}$ liters
74. 19
75. 1260
76. $-\sqrt{2}$
77. $\frac{5}{2}$ or 2.5
78. 9 sec
79. $\frac{x-y}{y}$
80. k^2
81. $12a - 6b$
82. 214
83. 45
84. $\frac{25}{6}$
85. 343
86. 4
87. 10 years old
88. 4 dollars
89. $\frac{8}{5}$ or 1.6
90. 7
91. 32 feet/sec
92. 362
93. 72,000 dollars
94. 61
95. 3 hours and 30 minutes
96. -22
97. 37
98. 18%
99. $\frac{7}{39}$
100. 24 days
101. 155
102. $46\frac{2}{13}$ %
103. $\frac{17}{27}$
104. 525
105. 1120
106. 26
107. $40
108. 210
109. 1 Chem book, 5 Phys books
110. $\frac{37}{64}$
111. 15
112. 12
113. 36
114. 362,880
115. 9
116. 24,024
117. 36°
118. .6
119. $\frac{1}{4}$ or .25
120. 37 dollars
121. 3
122. 39
123. 2664
124. $\frac{509}{5}$ or 101.8
125. $\frac{1}{6}$
126. 4
127. $\frac{11}{40}$
128. $\frac{331}{435}$
129. 26
130. 40
131. 47
132. 14
133. 35
134. $\frac{9}{35}$
135. $\frac{46}{165}$

ACTIONS TO COMPLETE AFTER YOU HAVE READ THIS BOOK

1. **Take a previously administered Mathcounts exam**
   You can find past exams on the Mathcounts website.

2. **Continue to practice a few problems each day**
   The free PDFs included with this book will give you lots of additional problems for extra practice.

3. **Review this book**
   If this book helped you, please post your positive feedback on the site you purchased it from; e.g. Amazon, Barnes and Noble, etc.

4. **Claim your FREE bonuses**
   If you have not done so yet, visit the following webpage and enter your email address to receive additional problems with solutions and a free PDF version of *555 Geometry Problems for High School Students*.

www.thesatmathprep.com/McountX.html

About the Authors

Tayyip Oral graduated from Qafqaz University in Azerbaijan in 1998 with a Bachelor's degree in Engineering, and he received an MBA from the same university in 2010. Tayyip is an educator who has written several books related to math and intelligence questions. He has previously taught math and IQ classes at Baku Araz preparatory school for 13 years.

Serife Turan graduated from Mehmet Akif Ersoy University in 2008 with a B.A. in education , and she worked for three years as a math teacher. She is currently pursuing her Master's Degree in the School of Education and Human Development at the University of Houston Victory.

Dr. Steve Warner earned his Ph.D. at Rutgers University in Mathematics, and he currently works as an Associate Professor at Hofstra University. Dr. Warner has over 15 years of experience in general math tutoring and over 10 years of experience in SAT math tutoring. He has tutored students both individually and in group settings and has published several math prep books for the SAT, ACT and AP Calculus exams.

BOOKS BY TAYYIP ORAL

1. T. Oral and Dr. S. Warner, 555 Math IQ Questions for Middle School Students, USA, 2015

2. T. Oral and Sevket Oral, 555 Math IQ Questions for Elementary School Students, USA, 2015

3. T. Oral, IQ Intelligence Questions for Middle and High School Students, USA, 2014

4. T. Oral, E. Seyidzade, Araz publishing, Master's Degree Program Preparation (IQ), Cag Ogretim, Araz Courses, Baku, Azerbaijan, 2010, Azerbaijan.

5. T. oral,M. Aranli, F. Sadigov, and N. Resullu, A Text Book for Job Placement Exam in Azerbaijan for Undergraduate and Post Undergraduate Students in Azerbaijan, Resullu publishing, Baku, Azerbaijan - 2012 (3.edition)

6. T. Oral and I. Hesenov, Algebra (Text Book), Nurlar Printing and Publishing, Baku, Azerbaijan, 2001.

7. T. Oral, I. Hesenov, S. Maharramov, and J. Mikaylov, Geometry (Text Book), Nurlar Printing and Publishing, Baku, Azerbaijan, 2002.

8. T. Oral, I. Hesenov, and S. Maharramov, Geometry Formulas (Text Book), Araz courses, Baku, Azerbaijan, 2003.

9. T. Oral, I. Hesenov, and S. Maharramov, Algebra Formulas (Text Book), Araz courses, Baku, Azerbaijan, 2000

BOOKS BY DR. STEVE WARNER

LEVEL 1
320
SAT MATH
SUBJECT TEST
PROBLEMS
By Dr. Steve Warner

LEVEL 2
320
SAT MATH
SUBJECT TEST
PROBLEMS
Arranged by Topic
and Difficulty Level
By Dr. Steve Warner

GET 800
320
ACT MATH
PROBLEMS
By Dr. Steve Warner

GET 800
320
AP CALCULUS
AB PROBLEMS
Arranged by Topic
and Difficulty Level
By Dr. Steve Warner

GET 800
320
AP CALCULUS
BC PROBLEMS
Arranged by Topic
and Difficulty Level
By Dr. Steve Warner

GET 800
SAT
Prep Math
Survival
Guide
By Dr. Steve Warner
Essential Information For Students
That Want to Maximize Their Score
in SAT Math with Minimal Effort

555 Math IQ Questions
for Middle School Students
Dr. Steve Warner
Tayyip Oral

555 Geometry Problems
for High School Students
Dr. Steve Warner
Tayyip Oral
Serife Turan

Made in the USA
Middletown, DE
11 March 2018